Springer Texts in Business and Economics

Springer Texts in Business and Economics (STBE) delivers high-quality instructional content for undergraduates and graduates in all areas of Business/Management Science and Economics. The series is comprised of self-contained books with a broad and comprehensive coverage that are suitable for class as well as for individual self-study. All texts are authored by established experts in their fields and offer a solid methodological background, often accompanied by problems and exercises.

Anja Blatter • Sean Bradbury • Pascal Bruhn • Dietmar Ernst

Risk Management in Banks and Insurance Companies

Step by Step

Anja Blatter
Nürtingen-Geislingen University
of Applied Sciences
Nürtingen, Germany

Sean Bradbury
Nürtingen-Geislingen University
of Applied Sciences
Nürtingen, Germany

Pascal Bruhn
Nürtingen-Geislingen University
of Applied Sciences
Nürtingen, Germany

Dietmar Ernst
Nürtingen-Geislingen University
of Applied Sciences
Nürtingen, Germany

ISSN 2192-4333 ISSN 2192-4341 (electronic)
Springer Texts in Business and Economics
ISBN 978-3-031-42838-8 ISBN 978-3-031-42836-4 (eBook)
https://doi.org/10.1007/978-3-031-42836-4

This work contains media enhancements, which are displayed with a "play" icon. Material in the print book can be viewed on a mobile device by downloading the Springer Nature "More Media" app available in the major app stores. The media enhancements in the online version of the work can be accessed directly by authorized users.

Translation from the German language edition: "Risikomanagement bei Banken und Versicherungen Schritt für Schritt" by Anja Blatter et al., © UTB GmbH 2023. Published by UTB GmbH. All Rights Reserved.

© The Editor(s) (if applicable) and The Author(s), under exclusive license to Springer Nature Switzerland AG 2024
This work is subject to copyright. All rights are solely and exclusively licensed by the Publisher, whether the whole or part of the material is concerned, specifically the rights of reprinting, reuse of illustrations, recitation, broadcasting, reproduction on microfilms or in any other physical way, and transmission or information storage and retrieval, electronic adaptation, computer software, or by similar or dissimilar methodology now known or hereafter developed.
The use of general descriptive names, registered names, trademarks, service marks, etc. in this publication does not imply, even in the absence of a specific statement, that such names are exempt from the relevant protective laws and regulations and therefore free for general use.
The publisher, the authors, and the editors are safe to assume that the advice and information in this book are believed to be true and accurate at the date of publication. Neither the publisher nor the authors or the editors give a warranty, expressed or implied, with respect to the material contained herein or for any errors or omissions that may have been made. The publisher remains neutral with regard to jurisdictional claims in published maps and institutional affiliations.

This Springer imprint is published by the registered company Springer Nature Switzerland AG
The registered company address is: Gewerbestrasse 11, 6330 Cham, Switzerland

If disposing of this product, please recycle the paper.

Contents

1	**Introduction**	1
	Risks and Their Sources	1
	What Is Risk?	2
	Structure of the Book	2
2	**Market Risks**	5
	Course Unit 1: Return and Volatility	5
	Assignment 1: Return Calculation	5
	Assignment 2: Creating a Histogram	11
	Assignment 3: Creation of a Density Function and a Distribution Function	16
	Assignment 4: Calculation of the Variance	24
	Assignment 5: Calculation of the Standard Deviation	27
	Course Unit 2: Modelling Volatilities	30
	Assignment 6: Calculation of Volatility with the EWMA Model	30
	Assignment 7: Calculation of Volatility with the ARCH Model	38
	Assignment 8: Calculation of Volatility with the GARCH Model	45
	Course Unit 3: Modelling of Stochastic Processes	53
	Assignment 9: Geometric Brownian Motion	53
	Assignment 10: Vasicek/Ornstein-Uhlenbeck Process	61
	Course Unit 4: Derivation of Risk Ratios with the Help of Black-Scholes	68
	Assignment 11: From Geometric Brownian Motion to Black-Scholes	68
	Assignment 12: Excursus: Put-Call Parity	74
	Assignment 13: Risk Metrics: The Greeks	78
	Assignment 14: Implied Volatility—A Key Driver in Black-Scholes	83
	Assignment 15: Volatility-Smile/-Surface	86

3 Credit Risks ... 93
Assignment 16: Rating Migration Matrices ... 93
Assignment 17: Merton's Model ... 98
Assignment 18: Vasicek Model—Calculation of the Worst-Case Default Rate ... 103
Assignment 19: Vasicek Model—Simulation of the Annual Portfolio Default Rate ... 107
Assignment 20: Vasicek Model—Estimation of Parameters from Historical Data ... 112
Assignment 21: Vasicek Model—Calculation of the Portfolio Loss ... 115

4 Operational Risks ... 119
Assignment 22: Calibration of the Loss Distribution Based on Expert Judgement ... 119

5 Risk Measures ... 127
Course Unit 1: Value at Risk-Risk Measures ... 127
    Assignment 23: Calculating the Value at Risk for a Discrete Probability Distribution ... 127
    Assignment 24: Calculation of the Mean Value at Risk for a Discrete Probability Distribution ... 132
    Assignment 25: Calculation of the Conditional Value at Risk/Expected Shortfall/Tail Value at Risk for a Discrete Probability Distribution ... 135
    Assignment 26: Calculation of the Value at Risk with a Continuous Probability Distribution ... 138
    Assignment 27: Calculation of the Conditional Value at Risk or Expected Shortfall for a Continuous Probability Distribution ... 145
    Assignment 28: Backtesting—How Good is the Value at Risk? ... 149
Course Unit 2: Lower Partial Moment Risk Measures ... 154
    Assignment 29: Calculation of Lower Partial Moments—Shortfall Probability ... 154
    Assignment 30: Calculation of Lower Partial Moments—Shortfall Expectation Value ... 156
    Assignment 31: Calculation of Lower Partial Moments—Shortfall Variance ... 158
Course Unit 3: Bond Risk Measures, Extreme Risks and Risk Measures in Comparison ... 161
    Assignment 32: Macaulay Duration and Modified Duration ... 161
    Assignment 33: Extreme Value Theory ... 172
    Assignment 34: Risk Measures in Comparison ... 180

6 Aggregation ... 183
Assignment 35: Variance–Covariance Method: Variance–Covariance Matrix and Portfolio Risk ... 183

Assignment 36: Variance–Covariance Method: Calculation of the Value
at Risk and Conditional Value at Risk 188
Assignment 37: Generation of Copulas 192
Assignment 38: Modelling the Aggregated Risk Using Copulas 199
Assignment 39: Risk Capital 207

References 211

Index 213

Chapter 1
Introduction

Risks and Their Sources

Facing multi-faceted and complex challenges in the modern world, banks and insurance companies need up-to-date, flexible and efficient risk management tools that cover all relevant risk areas.

Risk is therefore one of the main challenges in finance. Any activity of a bank or insurance company, like any other field of economics, involves a certain degree of risk. Risks can arise, for example, when granting loans, withdrawing deposits, advising on financing, especially corporate financing, or investing money. In addition, management must ensure that employees comply with the company's rules, as financial risks can occur from legal disputes and reputational damage. In this book, you will learn to quantify, simulate and hedge financial risks. Every bank and insurance company that provides a financial service is required by law to control and manage risks within the framework of quantitative risk management.

There are several risks that a financial services provider has to consider, the first of which is *market risk*. Market risks include an institution's foreign currency—and commodity risks, as well as position risks (interest rate and equity price-related risks) of the trading book. Market risk describes the risk that the market value of the portfolio will deviate from the expected value.

The second risk to consider is *credit risk*. It describes the risk that a borrower will not meet the contractually agreed interest and principal payments. It also includes deteriorations in the creditworthiness of business partners, for example in the form of rating downgrades. It is important to note that in the event of default, part of the capital invested is reimbursed either from the insolvency estate or from guarantees. Therefore, a crucial component of credit risk is the determination of this amount.

The third risk in financial institutions is *operational risk*. Typically operational risk in banks and insurances is the risk of loss resulting from inadequate or failed internal processes, people and systems or from external events. These risks can be exceptionally devastating as financial institutions rely very heavily on their reputation to gain trust with their clients.

Lastly further risks may include insurance risks and liquidity risks, which will not be highlighted as part of this book. However, some of the methods described in this book can also be applied to these risks.

What Is Risk?

Every person has his own personal concept of "risks". For the vast majority, it is some form of existential threat, such as extreme weather, being laid-off or some sort of devastating natural disaster such as earthquakes, floods or draughts. Since everyone has a personal definition of "risk" or a different tolerance level for "risk", it is important to specify what constitutes as a "risk" in the confines of this book.

For the purpose of this book we will rely on the risk definition used in financial literature. Usually in this specific field, risk is defined as a deviation from the expected value, which in its broadest interpretation includes both a positive and a negative deviation from the expected value. However, positive deviations do not require monitoring and as such are not included as risk in this book. Thus, for the purpose of this book, a "risk" is defined as a financially negative deviation from the expected value.

Structure of the Book

This book is structured by the individual components of financial risk mentioned in the introduction. Every chapter will introduce a different risk component of a financial institution.

In Chap. 2, you will learn the basics for analysing and modelling market risks. Many of the methods and concepts introduced in Chap. 2 will also be applied in the following course on credit risks. The first section revolving around market risks deals with the calculation of returns, the distribution function, variance and standard deviation or volatility. After this methodical introduction, you will learn to calculate interest rates using deterministic models. The central models of this segment are ARCH models and specifically the generalised GARCH model, which all assume a clustering of volatility in the data. In the following assignment, stock prices are modelled with the help of stochastic models. Central to this segment is the Geometric Brownian Motion. In addition, option pricing will be touched on, since this is crucial for determining Implied Volatility. The corresponding results from the models can be used to determine risk measures using the concepts from Chap. 4.

Chapter 3 introduces the models to calculate credit risks. One possible method we will use is the usage of rating migration matrices, which contain the probabilities of rating transitions. The next assignment will take a closer look at interest rate risks using Vasicek models, as these constitute a crucial component of credit risk. The chapter is concluded by the determination of the Probability of Default using Merton's model.

In Chap. 4, operational risks are quantified by calibrating loss distributions based on expert estimates.

Chapter 5 takes a closer look at individual risk measures. The central risk measure from a regulatory perspective is the Value at Risk, as it determines the risk capital, i.e. the amount of capital that an institution must reserve for rare adverse events. In the course of this unit, other risk measures will be discussed such as Conditional Value at Risk and Lower Partial Moments. In addition, these risk measures are applied in Extreme Value Theory. Furthermore, the bond- and credit-specific risk measure Duration is presented.

In order to calculate a risk measure for an overall portfolio to determine the risk capital of a financial institution, the possible methods for aggregations have to be considered. There are various popular concepts for this case, which will be presented in Chap. 6. First, the concept of the variance–covariance matrix is explained. After that the possibility of presenting the implementation of Copulas is introduced. Copulas have the advantage that they can also represent correlations in extreme situations—a phenomenon that can often be observed in financial practice and is therefore highly relevant. Based on these dependencies, a new aggregate risk capital can be determined.

This book is designed to enable you to grasp and quantify the financial risks of a financial services provider in their entirety and to determine the risk capital required by the supervisory board. It presents the basis of the Certified Financial Engineer (CFE) certificate course in risk management for banks and insurance companies. For more information, visit www.certified-financial-engineer.de.

At this point, we would also like to express our sincere thanks to all those who provided us with professional support during the preparation of the book. Many thanks for the openness to follow new didactic paths.

We are very happy to receive questions and suggestions about our book.

Prof. Dr. Anja Blatter, Sean Bradbury, Pascal Bruhn and Prof. Dr. Dr. Dietmar Ernst.

Detailed Structure of the Case Study
Chapter 2: Market risks
   Course Unit 1: Return and Volatility

- You will learn how to calculate discrete and continuous returns.
- You will be able to graph discrete and continuous returns and explain the statistical concepts behind them.
- You will learn how to calculate the standard deviation and variance.

   Course Unit 2: Modelling Volatility

(continued)

- You will learn how to model the change in returns using the ARCH model.
- You will learn how to model change in returns using the GARCH model.

Course Unit 3: Modelling of Stock Prices

- You will master and understand the different methods to simulate stock prices.
- You will be able to model various stochastic processes in Matlab.
- You will learn the assumptions and calculation of the Black-Scholes model.
- You can apply the Black-Scholes model to calculate option prices as well as implied volatilities.
- You understand the modelling issues involved in calculating the Black-Scholes formula.

Chapter 3: Credit Risks

- You will learn to determine rating shifts in a specific time frame using rating migration matrices.
- You will learn how to apply Merton's model to determine a company's Probability of Default.

Chapter 4: Operational Risks

- You will be able to calibrate a loss distribution based on an expert estimate.

Chapter 5: Risk Measures

- You will learn about different measures of risk.
- You will learn how to apply Value at Risk and Conditional Value at Risk in the continuous and discrete cases.
- You can check the accuracy of the risk measure.
- You will learn the specific risk measures for bonds.
- You will get to know the advantages and disadvantages of risk measures, especially whether a risk measure is coherent or not.

Chapter 6: Aggregation

- You will learn to create a variance–covariance matrix.
- You know what a Copula is and how to model it in Matlab.
- You can create a risk measure for a portfolio with different Copulas.

Chapter 2
Market Risks

Course Unit 1: Return and Volatility

Assignment 1: Return Calculation

Task

Calculate the discrete daily return and the continuous daily return for the MSCI WORLD index price from 12/31 t(5) retrospectively for the last 5 years.

Content

> In this book, we will look at risks arising from assets such as commodities, exchange rates and interest rates. The definition of risk is based on the risk management perspective. Thus, *risk* is understood as a deviation from the expected value. In its broadest definition, this includes both a positive and a negative deviation from the expected value. However, positive deviations do not require monitoring and are not considered as a risk by this book's definition. Thus, for the purpose of this book, a risk is a financially negative deviation from the expected value.
>
> Risks can occur from changes in prices or values for assets. These can be measured in *absolute terms* (the value of the stock has increased by 5.00 USD) or in *relative terms* (the value of the stock has increased by 5.0%). Using

(continued)

Supplementary Information The online version contains supplementary material available at https://doi.org/10.1007/978-3-031-42836-4_2.

relative changes allows risks from different assets and asset classes to be compared and aggregated into an overall risk. The relative changes in value are referred to as interest rates for interest-bearing financial products and yields for other financial products. In the following, the uniform term *return* is used. It is possible to distinguish between

- discrete returns
- continuous returns.

The relative change in value or *discrete return* r^d considers two individual points in time (investment time and the end of the investment period) or several investment points in time within an investment period.

In the case of a *continuous return* r^s it is assumed that the capital invested earns interest on a continuous basis. The difference between continuous and discrete returns is described by the consideration of the periods in which the investment earns interest. It may well be that an investment earns interest not only monthly, but also weekly, daily or even hourly, or even at shorter intervals. With a continuous return, infinitesimally small investment periods are assumed. The smaller the interest periods are, the smaller the difference between the discrete and the continuous return.

Important Formulas

Workbook: Case Study Risk Management Worksheet: Returns

Calculation of *discrete daily return*:

$$r_t^d = \frac{V_t - V_{t-1}}{V_{t-1}} = \frac{V_t}{V_{t-1}} - 1 \tag{2.1}$$

r_t^d = Discrete return at time t
V_t = Value at the time t, here on the day t
V_{t-1} = Value at time $t-1$

Excel example: D8=C8/C7-1

Calculation of the continuous daily rate of return:

$$r^c = ln\left(\frac{V_t}{V_{t-1}}\right) \tag{2.2}$$

r^c = Continuous return
V_t = Value at the time t, here on the day t
V_{t-1} = Value at time $t-1$

Excel example: E8=LN(C8/C7)

Execution in Excel
- Create a column for the MSCI WORLD index price (column C). Link the cells in this column to the values from the MSCI WORLD Assumptions worksheet so that the MSCI WORLD index prices for the specified period are displayed on the Returns worksheet.
- Calculate the discrete daily return according to the above formula D8=C8/C7-1.
- Then calculate the continuous daily return according to the formula E8=LN(C8/C7) above.

Excel Results (Fig. 2.1)

	A	B	C	D	E
1					
2		Asset	MSCI World		
3					
4		Currency	in USD		
5					
6		Date	Price MSCI World	Discrete Returns	Continuous Returns
7		01.01.t(1)	1,688.66		
8		02.01.t(1)	1,665.22	-1.39%	-1.40%
9		03.01.t(1)	1,645.64	-1.18%	-1.18%
10		06.01.t(1)	1,630.45	-0.92%	-0.93%
11		07.01.t(1)	1,633.80	0.21%	0.21%
12		08.01.t(1)	1,628.13	-0.35%	-0.35%
13		09.01.t(1)	1,635.78	0.47%	0.47%
14		10.01.t(1)	1,664.14	1.73%	1.72%
15		13.01.t(1)	1,668.78	0.28%	0.28%
16		14.01.t(1)	1,667.91	-0.05%	-0.05%

Fig. 2.1 Discrete and continuous returns

In Matlab
- Import the prices of the MSCI World index as well as the corresponding points in time. To do this, under `Current Folder`, select the folder in which you saved the Excel file `Matlab Data`.

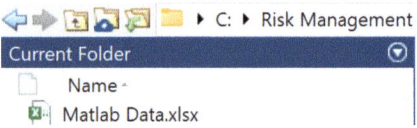

- Create a new `Live Script`.

- Enter the code below in the input lines. This imports the required data from the Excel file `Matlab Data`.

```
Data = readmatrix('Matlab Data.xlsx');
MSCI = Data(:,3);
Date = Data(:,1);
Price_MSCI = [Date,MSCI];
```

- Alternatively, you can also import data manually in Matlab. To do this, press the `Import Data` button on the `Home tab`.

- Then select the Excel document, mark columns `A` and `C` and import the data as a `Numeric Matrix`.

Course Unit 1: Return and Volatility

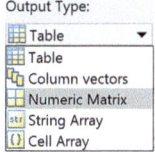

- Confirm the selection.

- Now calculate the discrete as well as the continuous returns of the MSCI World. Enter the following code in the input lines:

```
Discrete_Return = price2ret(Price_MSCI(:,2),[],'Periodic')
    Continuous_Return = price2ret(Price_MSCI(:,2))
```

Assignment 1: Return calculation

Import Database

Based on Excel File:

```
1    Data = readmatrix('Matlab Data.xlsx');
2    MSCI = Data(:,3);
3    Date = Data(:,1);
4    Price_MSCI = [Date,MSCI];
```

Computation of discrete/continous MSCI World returns

```
5    Discrete_Return = price2ret(Price_MSCI(:,2),[],'Periodic')
6    Continous_Return = price2ret(Price_MSCI(:,2))
```

- Press Run to start the script.

- Under Workspace you can view the imported data, defined variables as well as the calculated returns.

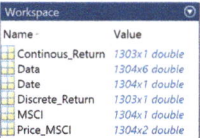

- Under Save you can name the Live Script and save it for later use.

Save

Matlab Results (Fig. 2.2)

	1 Date	2 Price MSCI	3 Discrete Return	4 Continous Return
1	42530	1.6887e+03	0	0
2	42531	1.6652e+03	-0.0139	-0.0140
3	42534	1.6456e+03	-0.0118	-0.0118
4	42535	1.6305e+03	-0.0092	-0.0093
5	42536	1.6338e+03	0.0021	0.0021
6	42537	1.6281e+03	-0.0035	-0.0035
7	42538	1.6358e+03	0.0047	0.0047
8	42541	1.6641e+03	0.0173	0.0172
9	42542	1.6688e+03	0.0028	0.0028
10	42543	1.6679e+03	-5.2134e-04	-5.2147e-04
11	42544	1.6918e+03	0.0143	0.0142
12	42545	1.6088e+03	-0.0490	-0.0503
13	42548	1.5718e+03	-0.0230	-0.0233
14	42549	1.5991e+03	0.0174	0.0173

Results — 1304x4 table

Fig. 2.2 Discrete and continuous returns in Matlab

Literature and Software References

- Ernst D., Häcker, J. (2017). *Financial Modeling - Business excellence in Financial Management, Corporate Finance, Portfolio Management and Derivatives*, Palgrave Macmillan. pp. 683–706.
- See Excel file: `Case Study Risk Management`, Excel worksheet: `Return`.
  See Matlab script: `A01_Return`.

Assignment 2: Creating a Histogram

Task

Create a histogram for the discrete daily returns of the MSCI WORLD index price from 12/31 t(5) retrospectively for the last 5 years to display the frequency distribution in a graph. Choose an appropriate division of the data into classes.

Content

A *histogram* is a graphical representation of the discrete frequency distribution of statistical data. It is a special form of the bar chart. The characteristic values are plotted on the X-axis and the frequencies on the Y-axis. The frequency of a measured value in a predefined interval is represented by a bar-shaped area above the interval—this can be relative (in percent) or absolute. In statistics, a histogram is called a frequency distribution.

A histogram provides a quick graphical overview of the distribution of returns. This opens up the possibility to illustrate the size of the spread and the risk of the asset. A histogram allows the user to visualise a larger amount of data more comprehensively than with a table, a concentration of extreme risks at the edges of the distribution can be identified easily.

Important Formulas

Workbook: `Case Study Risk Management` Worksheet: `Histogram`
  Determination of the *minimum* of the returns:

Excel example: H8=MIN(D8:D1310)

Determination of the maximum of the returns:

Excel example: H9=MAX(D8:D1310)

Determination of the mean value of the returns:

Excel example: H10=AVERAGE(D8:D1310)

Determination of the number of returns:

Excel example: H11=COUNT(D8:D1310)

Execution in Excel
- To create the histogram, we first calculate the discrete daily returns in column D.
- Then
  - the minimum H8=MIN(D8:D1310),
  - the maximum H9=MAX(D8:D1310),
  - the average value H10=AVERAGE(D8:D1310) as well as
  - the number of returns H11=COUNT(D8:D1310)
- are calculated with the respective Excel functions.
- This helps to define suitable intervals (class ranges) for the histogram.
- To create the histogram, specify the class range in cells F14:F40. The information is taken from the Assumptions worksheet.
- In our Excel example, the class range extends from −6.5% and increases to 6.5% in each 0.5% increment.
- By means of the analysis function HISTOGRAM the distribution can be determined very easily.
- You can access the function in Excel via Data; Data Analysis; Histogram. In case Data Analysis is not activated in your Excel version, go to File Options Add-Ins and check Analysis ToolPack and preferably also Solver Add-In, which we will need later. Confirm your selection with OK.

- Figure 2.3 shows the input area for creating the histogram.
- The discrete daily returns D8:D1310 are entered as the input range, the previously determined upper class limits in cells F14:F40 are entered as the class range, and cell G13, from which the result is to be displayed, is entered as the output range.
- Furthermore, the Chart Output field is clicked to immediately obtain a diagram from the data.
- The Class and Frequency columns are automatically inserted and calculated.

Course Unit 1: Return and Volatility

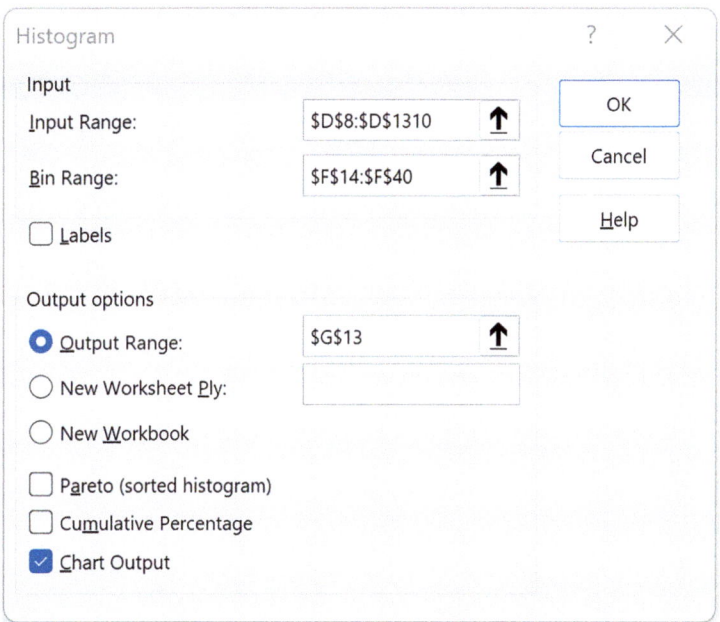

Fig. 2.3 Inputs for creating a histogram

- For optical reasons, it is recommended to adapt the chart in Excel to your own design afterwards.

Excel Results (Fig. 2.4)

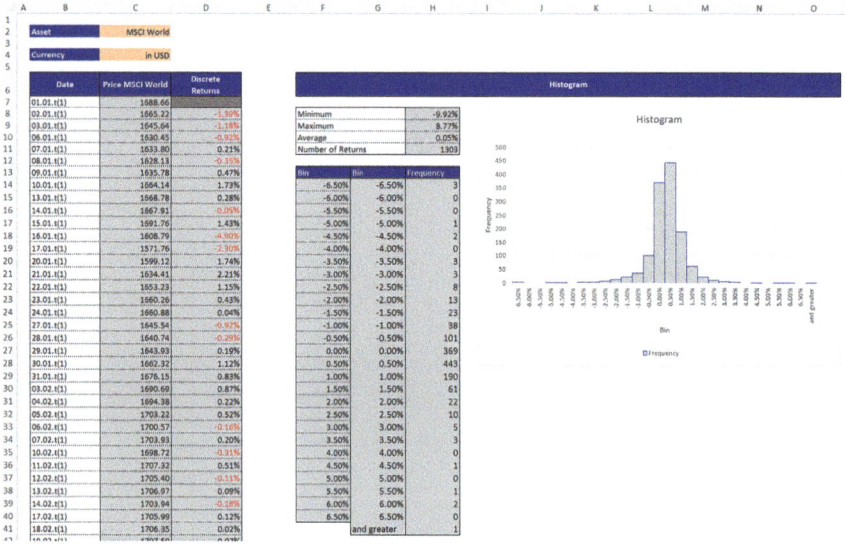

Fig. 2.4 Creating a histogram

Execution in Matlab
- Import the MSCI World prices and the associated dates.

```
Data = readmatrix('Matlab Data.xlsx');
MSCI = Data(:,3);
Date = Data(:,1);
Price_MSCI = [Date,MSCI];
```

- Calculate the discrete returns of the MSCI World.

```
Discrete_Return = price2ret(Price_MSCI(:,2),[],'Periodic')
```

- Describe the discrete returns.

```
Minimum = min(Discrete_Return)
Maximum = max(Discrete_Return)
Mean = mean(Discrete_Return)
Number_Observations = numel(Discrete_Return)
```

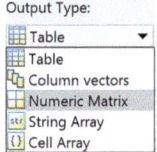

- Confirm the selection.

- Now calculate the discrete as well as the continuous returns of the MSCI World. Enter the following code in the input lines:

Discrete_Return = price2ret(Price_MSCI(:,2),[],'Periodic')
  Continuous_Return = price2ret(Price_MSCI(:,2))

Assignment 1: Return calculation

Import Database
Based on Excel File:

```
1  Data = readmatrix('Matlab Data.xlsx');
2  MSCI = Data(:,3);
3  Date = Data(:,1);
4  Price_MSCI = [Date,MSCI];
```

Computation of discrete/continous MSCI World returns

```
5  Discrete_Return = price2ret(Price_MSCI(:,2),[],'Periodic')
6  Continous_Return = price2ret(Price_MSCI(:,2))
```

- Press Run to start the script.

Run

- Under Workspace you can view the imported data, defined variables as well as the calculated returns.

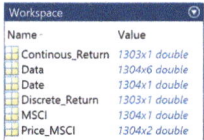

- Under Save you can name the Live Script and save it for later use.

Save

Matlab Results (Fig. 2.2)

	1 Date	2 Price MSCI	3 Discrete Return	4 Continous Return
1	42530	1.6887e+03	0	0
2	42531	1.6652e+03	-0.0139	-0.0140
3	42534	1.6456e+03	-0.0118	-0.0118
4	42535	1.6305e+03	-0.0092	-0.0093
5	42536	1.6338e+03	0.0021	0.0021
6	42537	1.6281e+03	-0.0035	-0.0035
7	42538	1.6358e+03	0.0047	0.0047
8	42541	1.6641e+03	0.0173	0.0172
9	42542	1.6688e+03	0.0028	0.0028
10	42543	1.6679e+03	-5.2134e-04	-5.2147e-04
11	42544	1.6918e+03	0.0143	0.0142
12	42545	1.6088e+03	-0.0490	-0.0503
13	42548	1.5718e+03	-0.0230	-0.0233
14	42549	1.5991e+03	0.0174	0.0173

Results 1304x4 table

Fig. 2.2 Discrete and continuous returns in Matlab

Literature and Software References

- Ernst D., Häcker, J. (2017). *Financial Modeling - Business excellence in Financial Management, Corporate Finance, Portfolio Management and Derivatives*, Palgrave Macmillan. pp. 683–706.
- See Excel file: `Case Study Risk Management`, Excel worksheet: `Return`.
  See Matlab script: `A01_Return`.

Assignment 2: Creating a Histogram

Task

Create a histogram for the discrete daily returns of the MSCI WORLD index price from 12/31 t(5) retrospectively for the last 5 years to display the frequency distribution in a graph. Choose an appropriate division of the data into classes.

Content

A *histogram* is a graphical representation of the discrete frequency distribution of statistical data. It is a special form of the bar chart. The characteristic values are plotted on the X-axis and the frequencies on the Y-axis. The frequency of a measured value in a predefined interval is represented by a bar-shaped area above the interval—this can be relative (in percent) or absolute. In statistics, a histogram is called a frequency distribution.

A histogram provides a quick graphical overview of the distribution of returns. This opens up the possibility to illustrate the size of the spread and the risk of the asset. A histogram allows the user to visualise a larger amount of data more comprehensively than with a table, a concentration of extreme risks at the edges of the distribution can be identified easily.

Important Formulas

Workbook: `Case Study Risk Management` Worksheet: `Histogram`
  Determination of the *minimum* of the returns:

Excel example: `H8=MIN(D8:D1310)`

Determination of the maximum of the returns:

> Excel example: H9=MAX(D8:D1310)

Determination of the mean value of the returns:

> Excel example: H10=AVERAGE(D8:D1310)

Determination of the number of returns:

> Excel example: H11=COUNT(D8:D1310)

Execution in Excel
- To create the histogram, we first calculate the discrete daily returns in column D.
- Then
  - the minimum H8=MIN(D8:D1310),
  - the maximum H9=MAX(D8:D1310),
  - the average value H10=AVERAGE(D8:D1310) as well as
  - the number of returns H11=COUNT(D8:D1310)
- are calculated with the respective Excel functions.
- This helps to define suitable intervals (class ranges) for the histogram.
- To create the histogram, specify the class range in cells F14:F40. The information is taken from the Assumptions worksheet.
- In our Excel example, the class range extends from −6.5% and increases to 6.5% in each 0.5% increment.
- By means of the analysis function HISTOGRAM the distribution can be determined very easily.
- You can access the function in Excel via Data; Data Analysis; Histogram. In case Data Analysis is not activated in your Excel version, go to File Options Add-Ins and check Analysis ToolPack and preferably also Solver Add-In, which we will need later. Confirm your selection with OK.

- Figure 2.3 shows the input area for creating the histogram.
- The discrete daily returns D8:D1310 are entered as the input range, the previously determined upper class limits in cells F14:F40 are entered as the class range, and cell G13, from which the result is to be displayed, is entered as the output range.
- Furthermore, the Chart Output field is clicked to immediately obtain a diagram from the data.
- The Class and Frequency columns are automatically inserted and calculated.

Course Unit 1: Return and Volatility

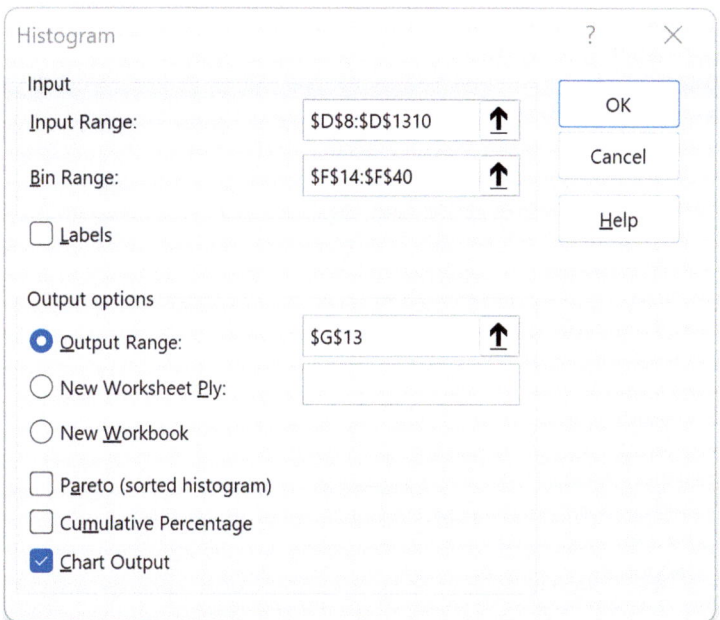

Fig. 2.3 Inputs for creating a histogram

- For optical reasons, it is recommended to adapt the chart in Excel to your own design afterwards.

Excel Results (Fig. 2.4)

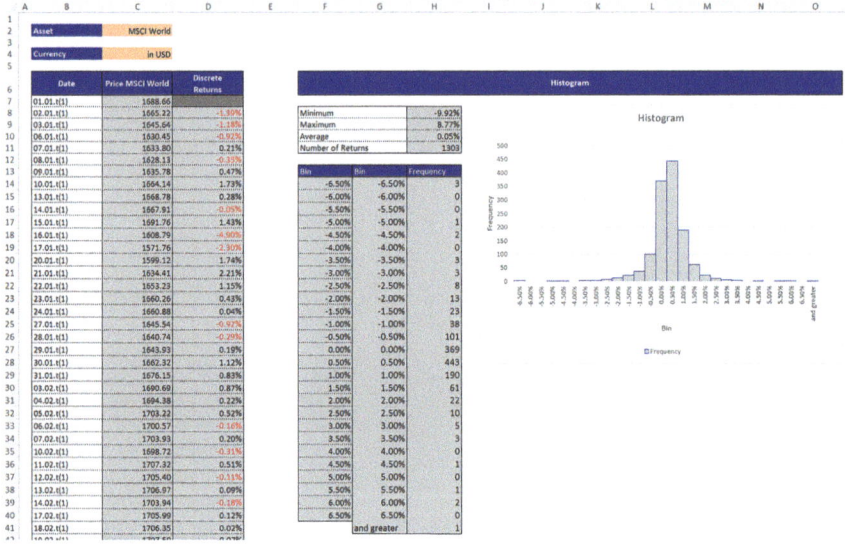

Fig. 2.4 Creating a histogram

Execution in Matlab
- Import the MSCI World prices and the associated dates.

```
Data = readmatrix('Matlab Data.xlsx');
MSCI = Data(:,3);
Date = Data(:,1);
Price_MSCI = [Date,MSCI];
```

- Calculate the discrete returns of the MSCI World.

```
Discrete_Return = price2ret(Price_MSCI(:,2),[],'Periodic')
```

- Describe the discrete returns.

```
Minimum = min(Discrete_Return)
Maximum = max(Discrete_Return)
Mean = mean(Discrete_Return)
Number_Observations = numel(Discrete_Return)
```

Course Unit 1: Return and Volatility

- Plot the returns on a histogram.

```
histogram(Discrete_Return)
   title 'MSCI World Returns';
```

Matlab Results (Fig. 2.5)

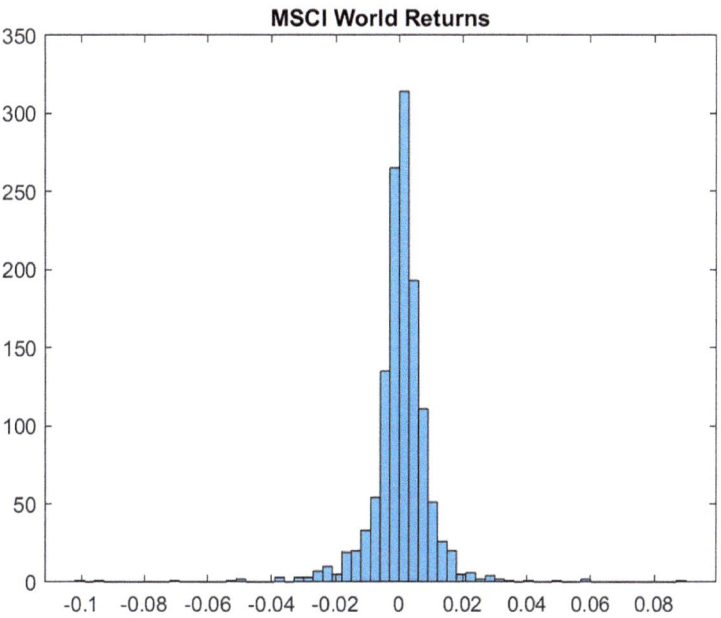

Fig. 2.5 Histogram of the MSCI World returns

Literature and Software References

> Ernst D., Häcker, J. (2017). *Financial Modeling - Business excellence in Financial Management, Corporate Finance, Portfolio Management and Derivatives*, Palgrave Macmillan. pp. 720–723.
>
> See Excel file: Case Study Risk Management, Excel worksheet: Histogram.
> See Matlab script: A02_Histogram.

Assignment 3: Creation of a Density Function and a Distribution Function

Task

Create a density function and a distribution function for the continuous daily returns of the MSCI WORLD index price based on the assumption of (a) a normal distribution and (b) a t-distribution, respectively.

Content

> For a *discrete* random variable, there is a finite amount of observed values, each of which has a positive probability. Therefore, probabilities can be assigned to the observed values. This assignment is called a probability function for discrete random variables. In the continuous case, on the other hand, the probability that a certain numerical value is taken is equal to 0. The reason for this is that continuous random variables assume real values, i.e. the precision can be constantly refined. Therefore, the density function replaces the probability function. The density function describes the probability that any value in an interval, i.e. a continuous set of values occur. The higher the density at a point, the higher the probability that a variable from this range is realised.
>
> A *distribution function* describes the relationship between a random variable and its probabilities. It indicates the probable outcomes of a random variable. For discrete random variables, the distribution function thus describes the cumulative probability. For continuous random variables, the distribution function can be interpreted as the area under the density function. The density function is therefore the first derivative of the distribution function.

(continued)

One of the most important continuous probability distributions is the *normal distribution*. The density function of the normal distribution is bell-shaped. The appearance and properties of the normal distribution are determined by two parameters:

- Expected value μ: It describes the number that the random variable assumes on average.
- Standard deviation σ: It shows the dispersion around the expected value.

The total area enclosed by the curve of the normal distribution (hence the integral from $-\infty$ to ∞), always has a value of one.

The *standard normal distribution* is a normal distribution in which the expected value $= 0$ and the standard deviation $= 1$.

Often the standard deviation is not known and must therefore be estimated. In this case, the usage of the *t-Student distribution* is recommended, which has a fatter tail-end than the normal distribution. This indicates that values further away from the expected value are assumed to be more probable compared to the normal distribution. The appearance and properties of the t-Student distribution is determined by one parameter:

- The *number of degrees of freedom n*. The larger the number of degrees of freedom, the closer the t-Student distribution approximates the standard normal distribution.

In addition to the normal distribution and the t-Student distribution, there are a large number of other distribution functions. To find out which distribution function best describes the data, the instrument of *calibration* is used. Statistical software or spreadsheet software is used to calibrate the empirical data to a theoretical distribution function. The distributions are fitted to the empirical data estimating the parameters for the distribution function. The software tools often also report the results of *goodness-of-fit tests* (for example, chi-square goodness-of-fit test, Box-Cox transformation, Kolmogorov-Smirnov goodness-of-fit test, Shapiro-Wilk test, or Anderson-Darling goodness-of-fit test). The estimated distribution parameters can then be used for risk assessment and modelling.

Important Formulas

Workbook: Case Study Risk Management Worksheet: Density & Distribution Function

Calculation of the *expected value*, which corresponds to the *mean value* of the continuous returns:

$$\mu = \frac{1}{m} \sum_{t=1}^{m} r_t \qquad (2.3)$$

μ = Expected value of the return
m = Number of observations
r_t = Return at time t

Excel example: E7=AVERAGE(Returns!E8:E1310)

Calculation of the *standard deviation* (starting from a sample) of the continuous daily returns:

$$\sigma = \sqrt{\frac{1}{m-1} \sum_{t=1}^{m} (r_t - \mu)^2} \qquad (2.4)$$

σ = Standard deviation of the return
m = Number of observations
r_t = Return at time t
μ = Expected value of the return

Excel example: E8=STDEV.S(Returns!E8:E1310)

Calculation of the *percentiles* of the normal distribution:

Excel example: C12=NORM.INV(B12,$E$7,$E$8)

Calculation of the *density* of the normal distribution:

$$f(t) = \frac{1}{\sigma\sqrt{2\pi}} \cdot e^{-\frac{1}{2}\left(\frac{t-\mu}{\sigma}\right)^2} \qquad (2.5)$$

σ = Standard deviation of the return
μ = Expected value of the return

Excel example: D12=NORM.DIST(C12,$E$7,$E$8,FALSE)

Calculation of the *distribution function* of the normal distribution:

$$F(t) = \frac{n}{\sigma\sqrt{2\pi}} \cdot \int_{-\infty}^{t} e^{-\frac{1}{2}\left(\frac{u-\mu}{\sigma}\right)^2} du \qquad (2.6)$$

Excel example: E12=NORM.DIST(C12,$E$7,$E$8,TRUE)

Calculation of the *percentiles* of the t-distribution:

Excel example: F12=T.INV(B12,$E$9)

Calculation of the *density* of the t-distribution:

Excel example: G12=T.DIST(F12,$E$9,FALSE)

Calculation of the *distribution function* of the t-distribution:

Excel example: H12=T.DIST(F12,$E$9,TRUE)

Execution in Excel

Note: In order to be able to represent a continuous distribution in Excel or Matlab, the values are often substituted by discrete values.

The reason for this is that it simplifies the calculation and less calculation capacity is required. This process is called discretisation.

- Create a column for the probabilities (column B). Link the cells in this column to the values from the Assumptions worksheet so that the probabilities are displayed on the Density & Distribution Function worksheet B12=Assumptions!C37.
- Calculate the mean of the returns in cell E7=AVERAGE(Returns!E8:E1310) and the standard deviation in E8=STDEV.S(Returns!E8:E1310). Assume 2 degrees of freedom for the t-distribution and enter them in cell E9.
- Based on this, calculate the percentiles of the normal distribution C12=NORM.INV(B12,$E$7,$E$8), the density D12=NORM.DIST(C12,$E$7,$E$8,FALSE) and the distribution function of the normal distribution according to the formulas above and E12=NORM.DIST(C12,$E$7,$E$8,TRUE).

- In the next step, determine the percentiles of the t-distribution, the density, and the distribution function of the t-distribution in the columns F, G and H. Assume degrees of freedom in the amount of 2. These result approximately, by the adjustment of a t-distribution to the empirical data. Such a fitting can be done, for example, in Matlab with the command fitdist.

Excel Results (Figs. 2.6, 2.7 and 2.8)

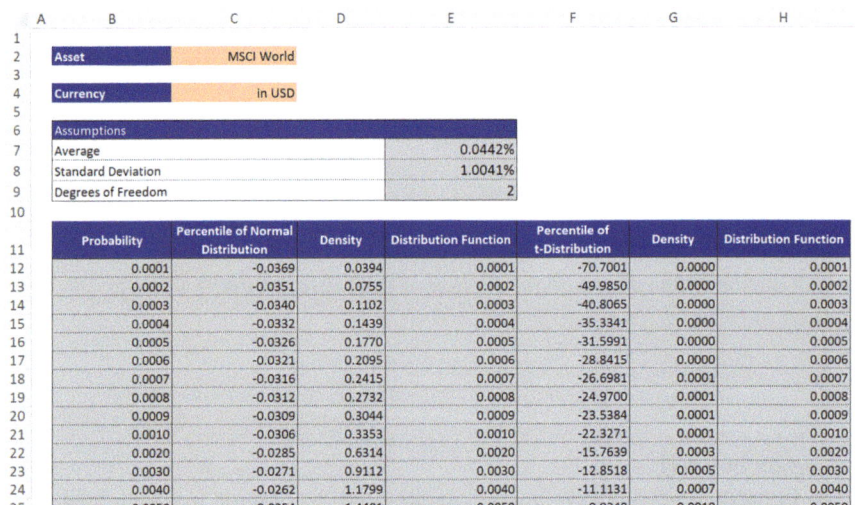

Fig. 2.6 Determination of a normal distribution and t-distribution in Excel

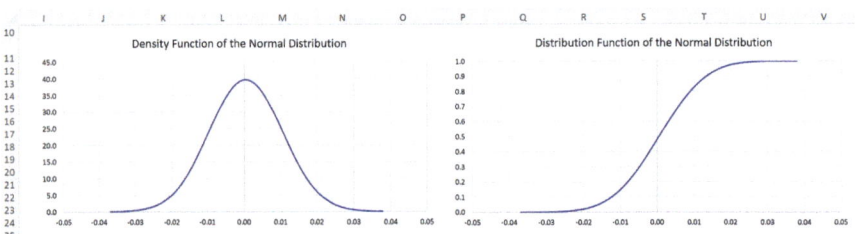

Fig. 2.7 Determination of a density and distribution function of the normal distribution in Excel

Course Unit 1: Return and Volatility

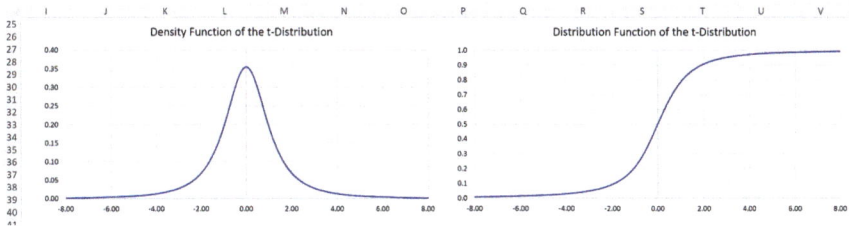

Fig. 2.8 Determination of a density and distribution function of the t-distribution in Excel

Execution in Matlab

- Import the MSCI World prices, the corresponding dates and the probability bins.

```
Data = readmatrix('Matlab Data.xlsx');
MSCI = Data(:,3);
Date = Data(:,1);
W = Data(:,4);
Probabilities = W(~isnan(Data(:,4))); % Assumed bins
Price_MSCI = [Date,MSCI];
```

- Calculate the continuous returns of the MSCI World.

```
Continuous_Return = price2ret(Price_MSCI(:,2));
```

- Describe the continuous returns.

```
Mean = mean(Continuous_Return)
Standard_Deviation = std(Continuous_Return)
```

- Create a normal distribution of the probabilities bins with mean and standard deviation of the continuous returns.

```
Normdist = norminv(Probabilities,Mean,Standard_Deviation);
```

- Create a density and distribution function based on the normal distribution.

```
Densityfct = normpdf(Normdist,Mean,Standard_Deviation);
Distributionfct = normcdf(Normdist,Mean,Standard_Deviation);
```

- Plot the functions.

```
figure
Standard = tiledlayout(1,2);
Standard.TileSpacing = 'compact';
title(Standard,'Standard Normal Distribution')
nexttile
plot(Normdist,Densityfct,'LineWidth',2)
title('Density function')
axis([-0.05 0.05 0 45])
nexttile
plot(Normdist,Distributionfct,'LineWidth',2)
title('Distribution function')
axis([−0.05 0.05 0 1])
```

- Determine the parameters of the t-distribution. The number of degrees of freedom can be determined with the `fitdist` command. To obtain consistent results with Excel, assume 2 degrees of freedom.

```
Parameter = fitdist(Continuous_Return,'tLocationScale');
Degrees_of_freedom = 2;
```

- Create a t-distribution.

```
t_dist = icdf('t',Probabilities,Degrees_of_freedom);
```

- Create a density and distribution function based on the t-distribution.

```
t_Densityfct = pdf('t',t_dist,Degrees_of_freedom);
t_Distributionfct = cdf('t',t_dist,Degrees_of_freedom);
```

- Plot the functions.

```
figure
tStudent = tiledlayout(1,2);
tStudent.TileSpacing = 'compact';
title(tStudent,'t-distribution')
nexttile
plot(t_dist,t_Densityfct,'LineWidth',2)
title('Density function')
axis([-10 10 0 0.4])
nexttile
plot(t_dist,t_Distributionfct,'LineWidth',2)
title('Distribution function')
axis([-10 10 0 1])
```

Matlab Results (Figs. 2.9 and 2.10)

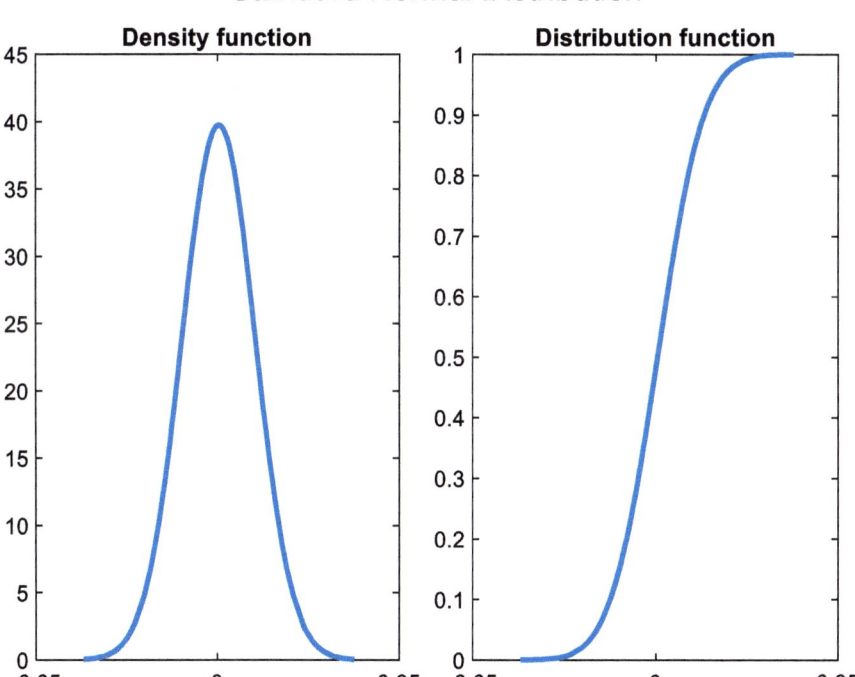

Fig. 2.9 The density and distribution function of the standard normal distribution

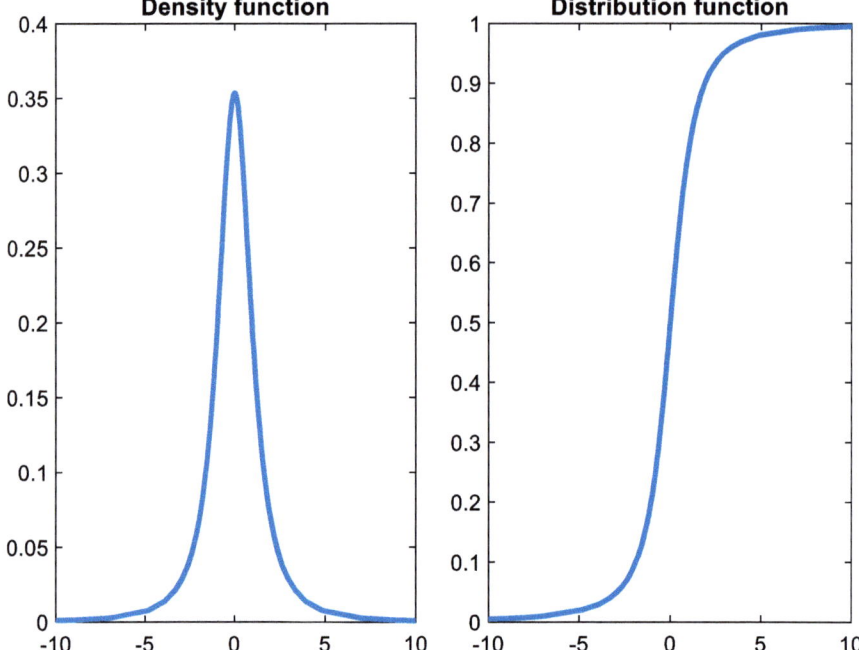

Fig. 2.10 The density and distribution function of the t-distribution

Literature and Software References

Ernst D., Häcker, J. (2017). *Financial Modeling - Business excellence in Financial Management, Corporate Finance, Portfolio Management and Derivatives*, Palgrave Macmillan. pp. 723–726.

See Excel file: Case Study Risk Management, Excel worksheet: Density & Distribution Function.
See Matlab script: A03_Density_Distribution_function.

Assignment 4: Calculation of the Variance

Task

Calculate the variance for the continuous daily returns of the MSCI WORLD index price based on a sample assumption. Also calculate the annualised and monthly variance.

Content

Variance measures the squared deviation of values around the mean.

Course Unit 1: Return and Volatility

Important Formulas
Workbook: `Case Study Risk Management` Worksheet: `Variance and Standard Deviation`

Calculation of the variance (starting from a sample) of the continuous daily returns:

$$\sigma^2 = \frac{1}{m-1} \sum_{t=1}^{m} (r_t - \mu)^2 \qquad (2.7)$$

σ^2 = Variance of returns
m = Number of observations
r_t = Return at time t
μ = Expected value of the return

> Excel example: `I8=VAR.S(D8:D1310)`

If the variance for a period of length t is given (e.g., the daily variance), then the following applies to the variance of a period comprising $m \cdot t$ time units (e.g., the annualised variance):

$$\sigma^2_{ann} = \sigma^2_t \cdot m \qquad (2.8)$$

$\sigma^2_{annualised}$ = Variance of return at annual level
σ^2_t = Variance of the return during the year
t = Single period during the year, e.g. day
m = Number of periods

> Excel example: `I9=I8*H9`

For the conversion of the variance of the return at an annual level into a variance of the return for a time frame within the year, the following applies:

$$\sigma^2_t = \sigma^2_{ann} \cdot \frac{1}{m} = \frac{\sigma^2_{ann}}{m} \qquad (2.9)$$

> Excel example: `I10=I9/H10`

Execution in Excel
- Calculate the variance based on the continuous daily returns in column D.
- Calculate the annualised and monthly variance based on the formulas above.

Excel Results (Fig. 2.11)

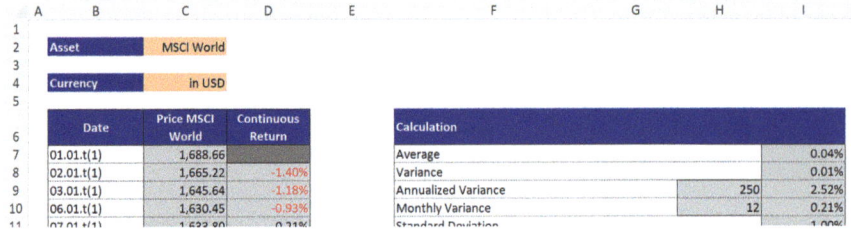

Fig. 2.11 Calculation of the variance

Execution in Matlab
- Import the MSCI World prices and the associated dates.

```
Data = readmatrix('Matlab Data.xlsx');
MSCI = Data(:,3);
Date = Data(:,1);
Price_MSCI = [Date,MSCI];
```

- Calculate the continuous returns of the MSCI World, as well as their mean.

```
Continuous_Return = price2ret(Price_MSCI(:,2));
Mean = mean(Continuous_Return)
```

- Calculate the variance of the continuous returns.

```
Daily_variance = var(Continuous_Return)
Annual_variance = Daily_variance*250
Monthly_variance = Annual_variance/12
```

Matlab Results (Fig. 2.12)

	Daily	Monthly	Annual
1 Variance	1.0082e-04	0.0021	0.0252

Fig. 2.12 Calculation of the variance in Matlab

Literature and Software References

> Ernst D., Häcker, J. (2017). *Financial Modeling - Business excellence in Financial Management, Corporate Finance, Portfolio Management and Derivatives*, Palgrave Macmillan, pp. 727–728.

> See Excel file: Case Study Risk Management, Excel worksheet: Variance and Standard Deviation.
> See Matlab script: A04_05_Variance_Standard_Deviation.

Assignment 5: Calculation of the Standard Deviation

Task

Calculate the standard deviation for the continuous daily returns of the MSCI WORLD index price based on a sample assumption. Also calculate the annualised and monthly standard deviation.

Content

> The *standard deviation* measures the deviation of values from the mean. The standard deviation is the square root of the variance.

Important Formulas

Workbook: Case Study Risk Management Worksheet: Variance and Standard Deviation

Calculation of the standard deviation (starting from a sample) of the continuous daily returns:

$$\sigma = \sqrt{\frac{1}{m-1} \sum_{t=1}^{m} (r_t - \mu)^2} \qquad (2.10)$$

σ = Standard deviation of the return
m = Number of observations

r_t = Return at time t
μ = Expected value of the return

Excel example: `I11=STDEV.S(D8:D1310)`

Calculation of the standard deviation (starting from a sample) as the root of the variance:

$$\sigma = \sqrt{\sigma^2} = \sqrt{Var[r]} \qquad (2.11)$$

σ = Standard deviation of the return
σ^2 = Variance

Excel example: `I12=SQRT(I8)`

The standard deviation is annualised as follows:

$$\sigma_{ann} = \sigma_t \cdot \sqrt{m} \qquad (2.12)$$

$\sigma_{annualised}$ = Standard deviation of return at annual level
σ_t = Standard deviation of return during the year
t = Single period during the year, e.g. day
m = Number of periods

Excel example: `I13=I11*SQRT(H13)`

The following applies to the conversion of the standard deviation of the return at the annual level into a standard deviation of the return within the year:

$$\sigma_t = \sigma_{ann} \cdot \frac{1}{\sqrt{m}} = \frac{\sigma_{ann}}{\sqrt{m}} \qquad (2.13)$$

Excel example: `I14=I13/SQRT(H14)`

Execution in Excel
- Calculate the standard deviation directly.
- First calculate the variance and then take the square root.
- Calculate the annualised and monthly standard deviation based on the formulas above.

		Daily	Monthly	Annual
1	Variance	1.0082e-04	0.0021	0.0252

Fig. 2.12 Calculation of the variance in Matlab

Literature and Software References

> Ernst D., Häcker, J. (2017). *Financial Modeling - Business excellence in Financial Management, Corporate Finance, Portfolio Management and Derivatives*, Palgrave Macmillan, pp. 727–728.
>
> See Excel file: `Case Study Risk Management`, Excel worksheet: `Variance and Standard Deviation`.
> See Matlab script: `A04_05_Variance_Standard_Deviation`.

Assignment 5: Calculation of the Standard Deviation

Task

Calculate the standard deviation for the continuous daily returns of the MSCI WORLD index price based on a sample assumption. Also calculate the annualised and monthly standard deviation.

Content

> The *standard deviation* measures the deviation of values from the mean. The standard deviation is the square root of the variance.

Important Formulas

Workbook: `Case Study Risk Management` Worksheet: `Variance and Standard Deviation`

Calculation of the standard deviation (starting from a sample) of the continuous daily returns:

$$\sigma = \sqrt{\frac{1}{m-1} \sum_{t=1}^{m} (r_t - \mu)^2} \qquad (2.10)$$

σ = Standard deviation of the return
m = Number of observations

r_t = Return at time t
μ = Expected value of the return

Excel example: `I11=STDEV.S(D8:D1310)`

Calculation of the standard deviation (starting from a sample) as the root of the variance:

$$\sigma = \sqrt{\sigma^2} = \sqrt{Var[r]} \qquad (2.11)$$

σ = Standard deviation of the return
σ^2 = Variance

Excel example: `I12=SQRT(I8)`

The standard deviation is annualised as follows:

$$\sigma_{ann} = \sigma_t \cdot \sqrt{m} \qquad (2.12)$$

$\sigma_{annualised}$ = Standard deviation of return at annual level
σ_t = Standard deviation of return during the year
t = Single period during the year, e.g. day
m = Number of periods

Excel example: `I13=I11*SQRT(H13)`

The following applies to the conversion of the standard deviation of the return at the annual level into a standard deviation of the return within the year:

$$\sigma_t = \sigma_{ann} \cdot \frac{1}{\sqrt{m}} = \frac{\sigma_{ann}}{\sqrt{m}} \qquad (2.13)$$

Excel example: `I14=I13/SQRT(H14)`

Execution in Excel
- Calculate the standard deviation directly.
- First calculate the variance and then take the square root.
- Calculate the annualised and monthly standard deviation based on the formulas above.

Course Unit 1: Return and Volatility

Excel Results (Fig. 2.13)

	A	B	C	D	E	F	G	H	I
11		07.01.t(1)	1,633.80	0.21%		Standard Deviation			1.00%
12		08.01.t(1)	1,628.13	-0.35%		Standard Deviation			1.00%
13		09.01.t(1)	1,635.78	0.47%		Annualized Standard Deviation		250	15.88%
14		10.01.t(1)	1,664.14	1.72%		Monthly Standard Deviation		12	4.58%
15		13.01.t(1)	1,668.70	0.28%					

Fig. 2.13 Calculation of the standard deviation

Execution in Matlab

- In continuation of Assignment 2.4, the standard deviation can be calculated easily. Note that Assignment 2.4 must precede the calculation of the standard deviation.
- Calculate the standard deviation of the continuous returns.

```
Standard_deviation = std(Continuous_Return)
    Annual_Standard_deviation = Standard_deviation*sqrt(250)
    Monthly_Standard_deviation = Annual_Standard_deviation/sqrt(12)
```

Matlab Results (Fig. 2.14)

		Daily	Monthly	Annual
1	Standard deviation	0.0100	0.0458	0.1588

Fig. 2.14 The calculated standard deviation in Matlab

Literature and Software References

> Ernst D., Häcker, J. (2017). *Financial Modeling - Business excellence in Financial Management, Corporate Finance, Portfolio Management and Derivatives*, Palgrave Macmillan, pp. 728–731.
>
> See Excel file: Case Study Risk Management, Excel worksheet: Variance and Standard Deviation.
> See Matlab script: A04_05_Variance_Standard_Deviation.

Course Unit 2: Modelling Volatilities

Assignment 6: Calculation of Volatility with the EWMA Model

Task

Calculate volatility using the EWMA model for the continuous daily returns of the MSCI World.

Content

> Estimates of volatility vary widely depending on the chosen time frames in the data basis. In order to improve the accuracy of the estimate of the variance or the standard deviation, statisticians usually favour as much data as possible. However, this also assumes a constant dispersion, i.e. a constant volatility over time. Accordingly, volatilities based on long periods of time indicate the extent to which a market typically moves over the long term. To capture the magnitude of ongoing variability, it is common to shorten the length of the observation period and calculate the standard deviation on a rolling basis for a fixed

(continued)

period of time. This shortened historical calculation base is then shifted further and further toward the current margin. This results in a time series whose individual elements are calculated as moving averages. This is referred to as *moving volatility*.

Since, when estimating current volatility, more recent values have a higher impact on the present than volatilities observed prior, we give a higher weight to more recent volatilities. To do this, we use exponentially decreasing weights. This leads to the exponentially weighted moving volatility (*EWMA*) *model*. In the EWMA model, historical values are exponentially weighted so that values in the near past are given a higher weight compared to older values. For this purpose, the weighting factor λ is used, which is also referred to as the lag factor or reduction factor. Its value is always between 0 and 1.

Within the EWMA model, the value of the weighting factor is λ is estimated such that the conditional EWMA variances are estimated close to the historical data. For this purpose, we use the *Maximum-Likelihood-Method*. The Maximum-Likelihood-Method calculates the value of the parameter λ that maximises the probability of occurrence of the historical values (here EWMA variance). The parameter λ is determined that it describes the observations that have occurred so far.

In practice, *two simplifications* have become established in the calculation of the exponentially weighted moving volatility, which we will retain here in the following calculations:

1. The arithmetic mean is set to zero when calculating the moving volatility on a daily basis. This assumption is justified by the fact that the expected change in the market price on a daily basis has no practical relevance.
2. $m-1$ is replaced by m. This change leads from the unbiased estimator to the maximum likelihood estimator which will play an important role for the following models.

Important Formulas

Workbook: Case Study Risk Management Worksheet: EWMA

The formula for the estimated value of the variance according to the EWMA model is:

$$\sigma_n^2 = \lambda \cdot \sigma_{n-1}^2 + (1-\lambda) \cdot r_{n-1}^2 \qquad (2.14)$$

λ = Degression rate
σ_{n-1}^2 = Estimated value for the variance of the previous day
r_{n-1}^2 = Actual rate of change of the previous day

The EWMA model shows that with only two values, the estimator for the previous day's variance σ_{n-1}^2 and the actual rate of change on the previous day

r_{n-1}^2 can be used to estimate the variance for the next day. Included in the estimate for the previous day's variance are historical values of daily change, which are accounted for with exponentially decreasing weights as they become more distant in the past. To demonstrate this, the resolution of the recursion formula is useful.

$$\sigma_n^2 = (1-\lambda) \cdot \sum (\lambda^{i-1} \cdot r_{n-i}^2 + \lambda^n \cdot \sigma_0^2) \text{ with } i = 1, \ldots, n \qquad (2.15)$$

The summand $\lambda^n \cdot \sigma_0^2$ converges to zero, since λ is regularly less than one and therefore λ^n converges to zero with increasing n. This leads to the following simplification:

$$\sigma_n^2 = (1-\lambda) \cdot \sum (\lambda^{i-1} \cdot r_{n-i}^2) \text{ with } i = 1, \ldots, n \qquad (2.16)$$

The weighting factor λ^{i-1} for the historical rates of change r_{n-i}^2 becomes smaller the larger i is, i.e. the further in the past the observation lies.

We now consider how the Maximum-Likelihood-Method can be used to estimate the parameter λ. Let r_i^2 be the observed variance and σ_i^2 the estimated variance. We assume that the probability distribution of the observed variances can be described as a normal distribution with expected value zero and variance σ_i^2.

The probability that the observations occur in the order in which they were observed is calculated using the following formula:

$$\prod_{i=1}^{m} \left[\frac{1}{\sqrt{2\pi\sigma_i^2}} exp\left(\frac{-r_i^2}{2\sigma_i^2}\right) \right] \qquad (2.17)$$

In the Maximum-Likelihood-Method, the value of σ_i^2 that maximises the expression above is the best estimator. Maximising an expression is equivalent to maximising the logarithm of the expression. By logarithmising, we can simplify the above formula as follows:

$$\sum_{i=1}^{m} \left(-ln(\sigma_i^2) - \frac{r_i^2}{\sigma_i^2} \right) \qquad (2.18)$$

Using the iterative search method (SOLVER in Excel) we can now determine the parameter λ which maximises the last listed formula.

Calculation of the daily variance:

Excel example: E12=D12^2

Calculation of the EWMA variance (initial value):

Excel example: F12=E12

Calculation of the EWMA variance (subsequent values):

Excel example: F13=F12*$D$7+(1-$D$7)*E12

Calculation of EWMA volatility:

Excel example: I12=SQRT(F12)

Calculation of the probability that the estimated value occurs:

Excel example: G12=-LN(F12)-E12/F12

Sum of probabilities that is maximised:

Excel example: G7=SUM(G12:G1314)

Initial value for λ which is used for optimisation in SOLVER:

Excel example: C7=Assumptions!C176

Execution in Excel
- Calculate the continuous daily returns in column D.
- In column E, based on the continuous daily returns, calculate the daily variance.
- The next step is to calculate the EWMA variance. For this we need the value for λ which we first obtain from the Assumptions. It is linked with cell C7. We link this cell again with the cell D7, in which after optimisation with the solver the value of λ resulting from the optimisation.
- For the calculation of the EWMA variance we need an initial value. Here we put in cell F12 the variance from cell E12.
- In F13 the actual formula for the EWMA variance is applied F13=F12*$D$7+(1-$D$7)*E12.
- The EWMA volatility is the root of the EWMA variance I12=SQRT(F12).
- Column G contains the calculation of the probability for the Maximum-Likelihood-Method. The Excel formula is G12=-LN(F12)-E12/F12.

- In cell G7, the probabilities are summed up G7=SUM(G12:G1314), which are then maximised during optimisation.
- For optimisation using the SOLVER, the following values must be entered into the SOLVER (Fig. 2.15).
- This results in a value for λ of 0.903918. This λ describes the historical observations of the EWMA variance best.
- If this value is used in the calculation of the moving volatility with exponentially decreasing weights the same values for the weighted variance result there as for the EWMA variance.

Excel Results (Fig. 2.16)

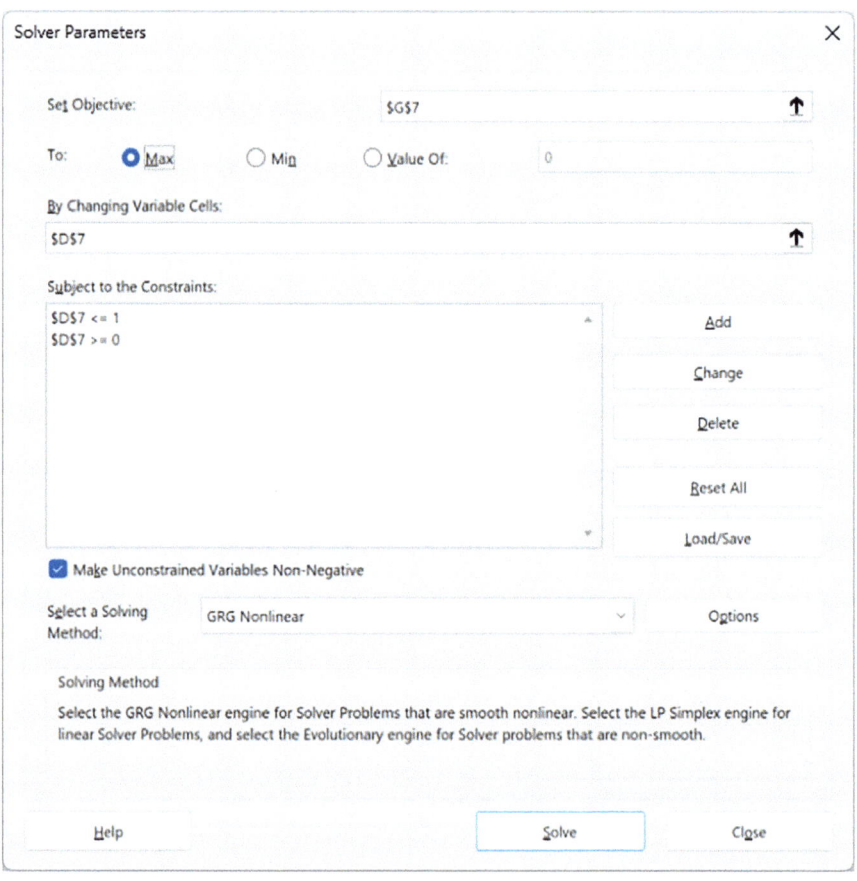

Fig. 2.15 Solver parameters for the EWMA model

Execution in Matlab

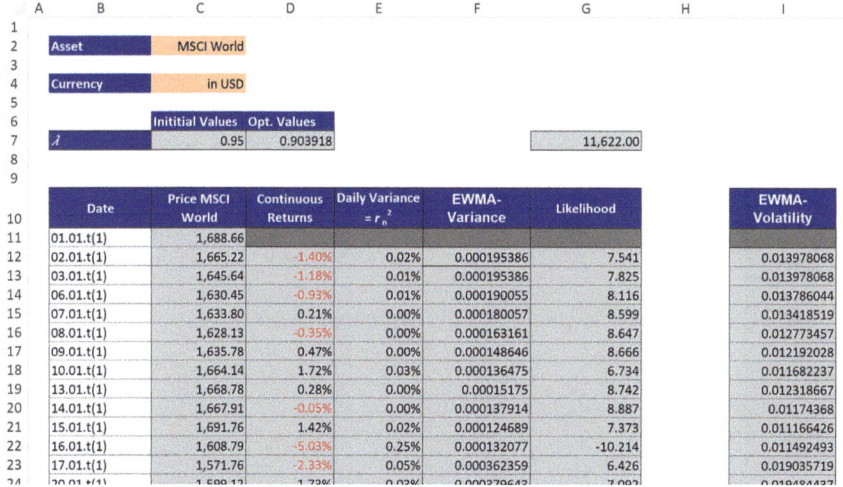

Fig. 2.16 Calculation of volatility with the EWMA model

- First define a function that calculates the estimated value of the variance, analogous to formula 1.2.6.1.

```
function [Sum_Likelihood] = EWMA_fun(Lambda)
global Daily_Variance;
y(1) = Daily_Variance(1);
for k = 2:length(Daily_Variance)
y(k) = y(k-1)*Lambda+(1-Lambda).*Daily_Variance(k-1);
end
Sum_Likelihood = -(sum(-log(y.')-Daily_Variance./(y.')));
    end
```

- Save the function as "EWMA_fun.m" and save it in the same folder as the code that follows.

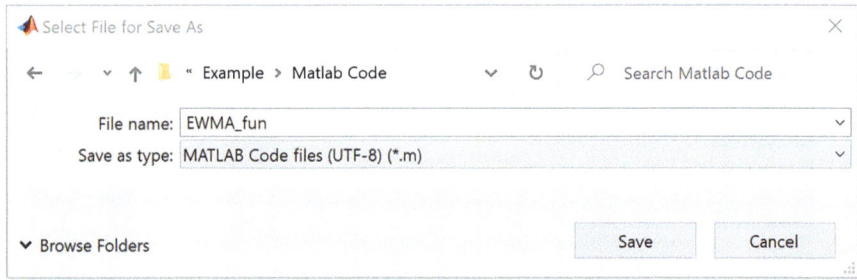

- Import the MSCI World prices and the associated dates.

  Data = readmatrix('Matlab Data.xlsx');
  MSCI = Data(:,3);
  Date = Data(:,1);
  Price_MSCI = [Date,MSCI];

- Define the global input variables.

  global Lambda;
  global Daily_Variance;

- Calculate the continuous returns of the MSCI World.

  Continuous_Return = price2ret(Price_MSCI(:,2));

- Determine the EWMA variance.

  format long
  Daily_Variance = Continuous_Return.^2; % In percent
  Lambda = 0.95;
  y(1) = Daily_Variance(1);
  for k = 2:length(Daily_Variance)
  y(k) = y(k-1)*Lambda+(1-Lambda).*Daily_Variance(k-1);
  end

(continued)

```
EWMA_Variance = y.';
Likelihood = -log(y.')-Daily_Variance./(y.');
EWMA_Vola = sqrt(EWMA_Variance);
format shortG
Sum_Likelihood = sum(Likelihood)
```

- Estimate the optimised weighting factor using the Maximum-Likelihood-Method.

```
fun = @EWMA_fun; % Call EWMA-function
[Lambda_Opt, SumLikelihood] = fminsearch(fun,Lambda); % Max-Likeli-Method
Likelihood_Opt = -(SumLikelihood);
Table = table(Lambda_Opt,Likelihood_Opt,'VariableNames',{'Lambda',
'Likelihood'},'RowNames',{'Optimized Values'})
```

Matlab Results (Fig. 2.17)

Fig. 2.17 Representation of the optimised lambda

		Lambda	Likelihood
1	Optimized Values	0.90389	11622

Literature and Software References

> Hull, J. (2018). *Risk Management and Financial Institutions*. 5th ed. Wiley. pp. 225–227.
> McNeil., A., Frey, R., Embrechts, P. (2015). *Quantitative Risk Management: Concepts, Techniques and Tools*, Princeton University Press, pp. 132–133.
> See Excel file: Case Study Risk Management, Excel worksheet: EWMA.
> See Matlab script: A06_EWMA.

Assignment 7: Calculation of Volatility with the ARCH Model

Task

Calculate volatility using the ARCH model for the continuous daily returns of the MSCI World.

Content

> Most theoretical approaches in capital market theory assume that the variance of returns is constant over time. However, looking more closely at rates of price change in speculative markets, it can be observed that while they fluctuate around a constant mean, their variability over time is not subject to constant fluctuations. Rather, there is a tendency toward volatility clustering. Volatility clustering describes the temporary concentration of high and low returns in absolute terms. This means that the development of volatility over time follows a certain pattern. A phase of high volatility is followed by a phase of low volatility and vice versa. Thus, volatility clusters are formed in each phase. Classical methods such as regression analysis or time series analysis assume a constant variance of forecast errors over time and are unsuitable to explain phenomena such as volatility clustering. Especially stock and exchange rates or interest rates exhibit such behavioural patterns, which require a nonlinear modelling approach. One such approach was developed by *Robert F. Engle,* who was awarded the Nobel Prize in Economics in 2003 for his work on ARCH models.

(continued)

Course Unit 2: Modelling Volatilities

The acronym *ARCH* stands for "Autoregressive Conditional Heteroscedasticity". ARCH models try to consider that volatilities follow a certain pattern. This property of changing volatilities over time is called *heteroscedasticity*. ARCH models are also *autoregressive*, i.e., volatility is measured as a function of its predecessor, i.e., it is *conditional*.

ARCH models are particularly well suited for short-term forecasts based on current historical data. Here, we consider data from the past five days for the ARCH model.

Important Formulas
Workbook: Case Study Risk Management Worksheet: ARCH
The formula for the variance σ^2 at time t in the ARCH(m) model is as follows:

$$\sigma_t^2 = \gamma \cdot V_L + \sum_{i=1}^{m} \alpha_i \cdot r_{n-1}^2 \qquad (2.19)$$

V_L = Long-term variance of the time series
γ = Weighting factor of V_L
r_{n-i}^2 = Squared return (= variance) on the previous day
α_i = Weighting factor on day i
m = Number of observations
n = Period of the estimator = $m+1$

The ARCH(m) model implies that the variance σ_t^2 can be explained using the average long-term variance and the weighted historical variance from m observations.
Since the sum of the weights equals one:

$$\gamma + \sum_{i=1}^{m} \alpha_i = 1 \qquad (2.20)$$

In the ARCH(m) model, younger observations are assigned higher weights and older observations are assigned lower weights.
With $\omega = \gamma \cdot V_L$ the above formula can also be written as follows:

$$\sigma_t^2 = \omega + \sum_{i=1}^{m} \alpha_i \cdot r_{n-1}^2 \tag{2.21}$$

Calculation of the daily variance:

Excel example: E12=D12^2

Initial value for ω which is used for optimisation in SOLVER:

Excel example: L3=Assumptions!C181

Initial value for $\alpha(1)$ which is used for optimisation in SOLVER:

Excel example: L4=Assumptions!C182

Calculation of the conditional variance:

Excel example: F17=$M$3+$M$4*E12+$M$5*E13+$M$6*E14+$M$7*E15+$M$8*E16

Calculation of the conditional volatility:

Excel example: I12=SQRT(F12)

Calculation of the probability that the estimated value occurs:

Excel example: G12=-LN(F12)-E12/F12

Sum of probabilities that is maximised:

Excel example: G8=SUM(G12:G1314)

Calculation of the weighting factor of γ_L:

Excel example: `I2=1-SUM(M4:M8)`

Calculation of the daily long-term variance V_L:

Excel example: `I3=M3/I2`

Calculation of daily long-term volatility:

Excel example: `I4=SQRT(I3)`

Calculation of annual long-term volatility:

Excel example: `I5=I4*SQRT(Assumptions!C187)`

Execution in Excel
- Calculate the continuous daily returns in column D.
- In column E, based on the continuous daily returns, calculate the daily variance.
- The next step is to calculate the conditional variance. For this we need for the variables ω and α_i output values, which we obtain from the Assumptions. They are linked to the cell L3 and L4:L8. We link this cell again with the cells M3 and M4:M8, in which after optimisation with the solver the values of λ and α_i will be estimated.
- In F17 is the actual formula for calculating the conditional variance, which is composed of the average long-term variance and 5 observations F17=$M$3+$M$4*E12+$M$5*E13+$M$6*E14+$M$7*E15+$M$8*E16.
- The conditional volatility is the root of the conditional variance I12=SQRT(F12).
- Column G contains the calculation of the probability for the Maximum-Likelihood-Method. The Excel formula is G12=-LN(F12)-E12/F12.
- In cell G8, the probabilities are summed up G8=SUM(G12:G1314), which are then maximised during optimisation.
- For optimisation using the SOLVER, the following values must be entered into the SOLVER (Fig. 2.18).

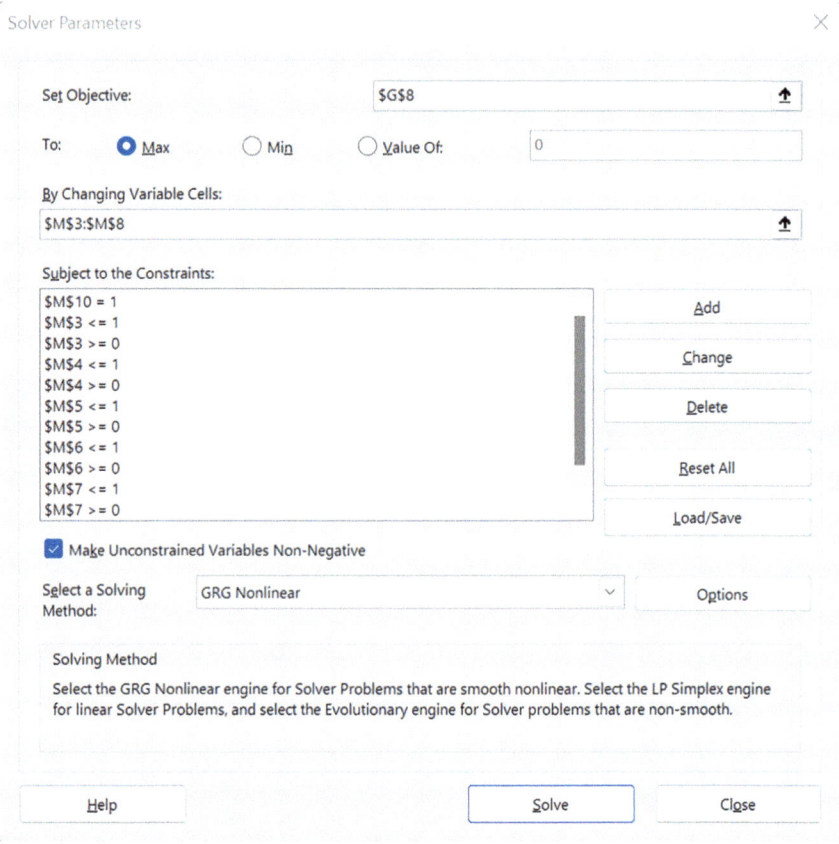

Fig. 2.18 Solver parameters for the ARCH model

- This results in a value for ω of 0.000013 and values for α_1 of 0.088174, for α_2 of 0.225761, for α_3 of 0.190161, for α_4 of 0.178966 and for α_5 of 0.252886. These values best describe the historical observations of the conditional variance.
- In the next step, we can now calculate the weighting factor γ_L `I2=1-SUM(M4:M8)`, the daily, long-term variance V_L `I3=M3/I2` and the daily, long-term volatility `I4=SQRT(I3)`.
- The result of the ARCH model used here states that the long-term volatility per year is 22.15%. The weights from the last five days of actually measured variances contribute 93.6% to this and the average, long-term variance of the time series is 6.4%.

Excel Results (Fig. 2.19)

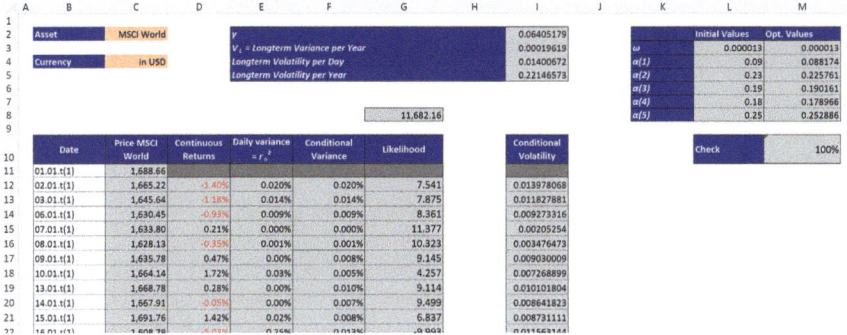

Fig. 2.19 Calculation of volatility with the ARCH model

Execution in Matlab
- Import the MSCI World prices and associated dates.

```
Data = readmatrix('Matlab Data.xlsx');
MSCI = Data(:,3);
Date = Data(:,1);
Price_MSCI = [Date,MSCI];
Tradings_days = 250;
```

- Calculate and plot the continuous returns of the MSCI World.

```
Continuous_Return = price2ret(Price_MSCI(:,2));
T = length(Continuous_Return);
figure
plot(Continuous_Return)
xlim([0, T])
title('MSCI World Returns')
ylabel('Returns')
xlabel('Days')
```

- Create an ARCH model with undefined parameters.

```
Model = garch('ARCHLags',[1 2 3 4 5]); % Takes into account the last 5 days
```

- Fit the model to the continuous returns and estimate the parameters.

```
% Estimation of parameters using Max-Likelihood-Method
estMdl = estimate(Model,Continuous_Return(6:end),'E0',Continuous_Return
(1:5));
Omega = estMdl.Constant % sprintf('%f',omega) for notation without e
Alpha5 = estMdl.ARCH{1} % Weighting factor of day i-1 (previous day)
Alpha4 = estMdl.ARCH{2} % ...
Alpha3 = estMdl.ARCH{3} % ...
Alpha2 = estMdl.ARCH{4} % ...
Alpha1 = estMdl.ARCH{5} % Weighting factor of day i-5
```

- Calculate the conditional variance as well as the conditional volatility, based on the determined parameters.

```
Daily_Variance = (Continuous_Return.^2)*100; % In percent -> following
values as well
Conditional_Variance = zeros(T,1);
Conditional_Variance(1:5) = Daily_Variance(1:5); % Initial values
for i = 6:T
Conditional_Variance(i) = Omega + Alpha1*Daily_Variance(i-5)+Alpha2*
Daily_Variance(i-4)+Alpha3*Daily_Variance(i-3)+Alpha4*
Daily_Variance(i-2)+Alpha5*Daily_Variance(i-1);
end
Conditional_Vola = sqrt(Conditional_Variance);
```

- Calculate the long-term variance and volatility per day and year.

```
Gamma = 1-(Alpha1+Alpha2+Alpha3+Alpha4+Alpha5);
% Weighting factor average longterm volatility
Variance_Day = Omega/Gamma; % Longterm variance per day
Vola_Day = sqrt(Variance_Day); % Longterm volatility per day
Vola_Year = Vola_Day*sqrt(Tradings_days); % Longterm volatility per year
Table   = table(Gamma,Variance_Day,Vola_Day,Vola_Year,'VariableNames',
{'Gamma', 'Daily Variance','Daily Volatility','Annual Volatility'})
```

Course Unit 2: Modelling Volatilities

Matlab Results (Figs. 2.20 and 2.21)

```
GARCH(0,5) Conditional Variance Model (Gaussian Distribution):
```

	Value	StandardError	TStatistic	PValue
Constant	1.2546e-05	1.0538e-06	11.905	1.1168e-32
ARCH{1}	0.25304	0.028683	8.8218	1.1268e-18
ARCH{2}	0.17906	0.028319	6.323	2.5645e-10
ARCH{3}	0.19021	0.033648	5.6529	1.5778e-08
ARCH{4}	0.22582	0.025488	8.86	8.0031e-19
ARCH{5}	0.088255	0.02113	4.1767	2.9579e-05

Fig. 2.20 The estimated ARCH model

	Gamma	Daily Variance	Daily Volatility	Annual Volatility
1	0.0636	1.9723e-04	0.0140	0.2221

Fig. 2.21 Representation of the calculated long-term variance and volatility

Literature and Software References

- McNeil., A., Frey, R., Embrechts, P. (2015). *Quantitative Risk Management: Concepts, Techniques and Tools*, Princeton University Press, pp. 112–118.
- See Excel file: Case Study Risk Management, Excel worksheet: ARCH.
  See Matlab script: A07_ARCH.

Assignment 8: Calculation of Volatility with the GARCH Model

Task

Calculate the volatility using the GARCH model for the continuous daily returns of the MSCI World.

Content

The idea of the ARCH models has been further developed in various ways and is now one of the advanced methods in econometrics. Its generalisations are the *GARCH models* (generalised autoregressive conditional heteroscedasticity), which were developed by *Bollerslev in* 1986. In contrast to the ARCH model, the conditional variance in GARCH models depends not only on the long-term variance and the weighted historical variance, but the conditional variance of the previous day is also considered.

GARCH models also allow volatility clusters to be considered when forecasting volatilities. This describes the property of volatilities to follow a certain pattern over time, which is called heteroskedasticity. Furthermore, autoregressive properties can be observed in volatility clusters, i.e. the volatility depends on its predecessor. In the presence of an autoregressive time series, the variance is conditional because its magnitude depends on the value of the previous variance. Simple GARCH models further assume that the underlying rates of change (returns) are normally distributed (but there are also GARCH models that do not assume a normal distribution of returns, e.g. EGARCH, GJR, etc.). However, their variance may vary over time and depends on the volatility of the previous periods. In this way, both the clustering of volatilities and a *leptokurtic distribution* can be modelled. Leptokurtic distributions show, on the one hand, fat tails which, in contrast to the normal distribution, make extreme price changes at the edges more likely and, on the other hand, they have more values around the expected value, i.e. the peak of the distribution is higher and narrower than in a normal distribution.

A challenge in the use of GARCH models is the estimation of the parameters. The estimation of the parameters is done with the help of the *Maximum-Likelihood-Method*. In this method, the parameters are estimated based on the historical values that they maximise the probability of the historical values. In other words, the parameters are determined that they best describe the observations that have occurred so far.

With a suitable choice of parameters, the GARCH(1,1) can be converted into the simpler ARCH model. For the case $= 0$, $=1 - \lambda$ and $=\lambda$ the ARCH model can be derived from the GARCH model, so that the latter can be regarded as a special case of GARCH(1,1).

Important Formulas

Workbook: Case Study Risk Management Worksheet: GARCH

The formula for the variance σ^2 at time t in the GARCH(1,1) model is:

Course Unit 2: Modelling Volatilities

$$\sigma_n^2 = \gamma \cdot V_L + \alpha \cdot r_{n-1}^2 + \beta \cdot \sigma_{n-1}^2 \qquad (2.22)$$

σ_n^2 = Conditional variance on current day n
V_L = Average, long-term variance of the time series
γ = Weighting factor of V_L
r_{n-1}^2 = Squared return actually measured on the previous day (= variance).
α = Weighting factor of r_{n-1}^2
σ_{n-1}^2 = Conditional variance of the previous day
β = Weighting factor of σ_{n-1}^2

The sum of the three weighting factors must equal one ($\gamma + \alpha + \beta = 1$).
Calculation of the daily variance:

Excel example: E12=D12^2

Initial value for ω which is used for optimisation in SOLVER:

Excel example: L3=Assumptions!C192

Initial value for α which is used for optimisation in SOLVER:

Excel example: L4=Assumptions!C193

Initial value for β which is used for optimisation in SOLVER:

Excel example: L5=Assumptions!C194

Calculation of the conditional variance (initial value):

Excel example: F12=E12

Calculation of the conditional variance (subsequent values):

Excel example: F13=$M$3+$M$4*E12+$M$5*F12

Calculation of the conditional volatility:

Excel example: I12=SQRT(F12)

Calculation of the probability that the estimated value occurs:

Excel example: G12=-LN(F12)-E12/F12

Sum of probabilities that is maximised:

Excel example: G8=SUM(G12:G1314)

Calculation of the weighting factor of γ:

Excel example: I2=1-M4-M5

Calculation of the daily long-term variance V_L:

Excel example: I3=M3/I2

Calculation of daily long-term volatility:

Excel example: I4=SQRT(I3)

Calculation of annual long-term volatility:

Excel example: I5=I4*SQRT(Assumptions!C195)

Execution in Excel
- Calculate the continuous daily returns in column D.
- In column E, based on the continuous daily returns, calculate the daily variance.
- The next step is to calculate the conditional variance. For this we need for the variables ω, α and β initial values, which we first obtain from the Assumptions. They are linked to the cells L3, L4 and L5. These values also form the output values for cells M3, M4 and M5. In cells M3, M4 and M5, after the optimisation with the solver, the values of ω, α and β will be in the cells M3, M4 and M5.

Course Unit 2: Modelling Volatilities

- For the calculation of the conditional variance we need an initial value. Here we insert the variance from cell E12 into cell F12.
- Then in F13 the actual calculation of the conditional variance takes place, which depends not only on the variance of the time series, but also on its own past, i.e. the conditional variance of the previous period F13=$M$3+$M$4*E12+$M$5*F12.
- The conditional volatility is the root of the conditional variance I12=SQRT(F12).
- Column G contains the calculation of the probability for the Maximum-Likelihood-Method. The Excel formula is G12=-LN(F12)-E12/F12.
- In cell G8, the probabilities are summed up G8=SUM(G12:G1314), which are then maximised during optimisation.
- For optimisation using the SOLVER, the following values must be entered into the SOLVER (Fig. 2.22).

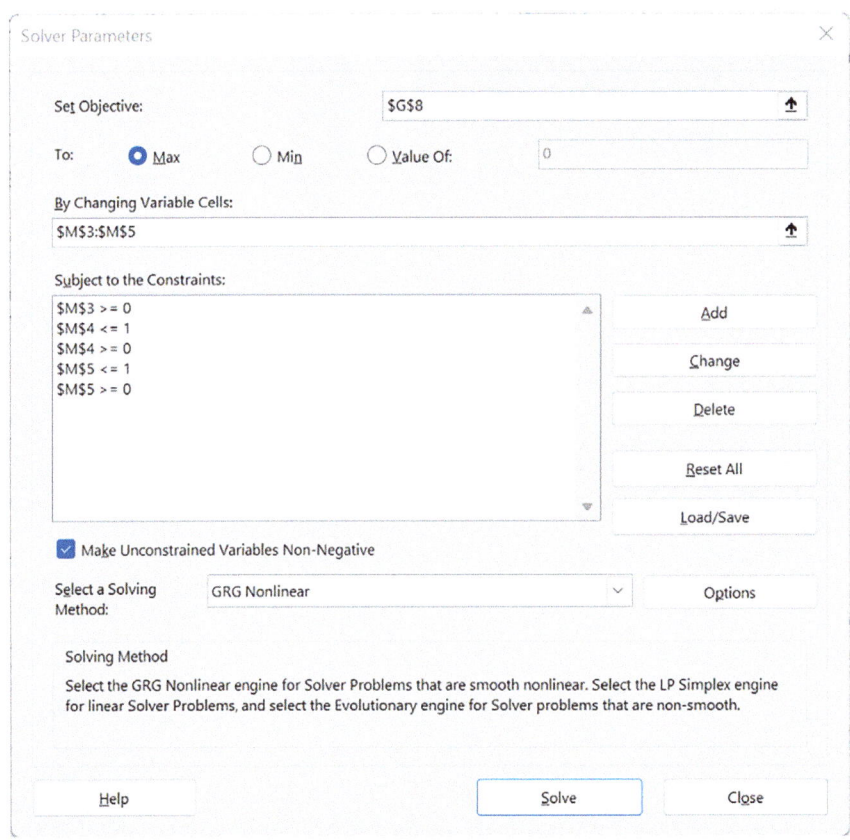

Fig. 2.22 Solver parameters for the GARCH model

- This results in values for ω in the amount of 0.0000025, for α a value of 0.23 and for β of 0.75.
- In the next step, the weighting factor of V_L `I2=1-M4-M5`, the daily, long-term variance V_L `I3=M3/I2` and the daily, long-term volatility `I4=SQRT(I3)` can be calculated. The long-term volatility per year is 16.96%.
- The result of the GARCH model used here states that the long-term volatility per year is 16.96%. The conditional variance of the previous day contributes 74.82%, the variance of the previous day 22.98% and the average, long-term variance of the time series 2.2%.

Excel Results (Fig. 2.23)

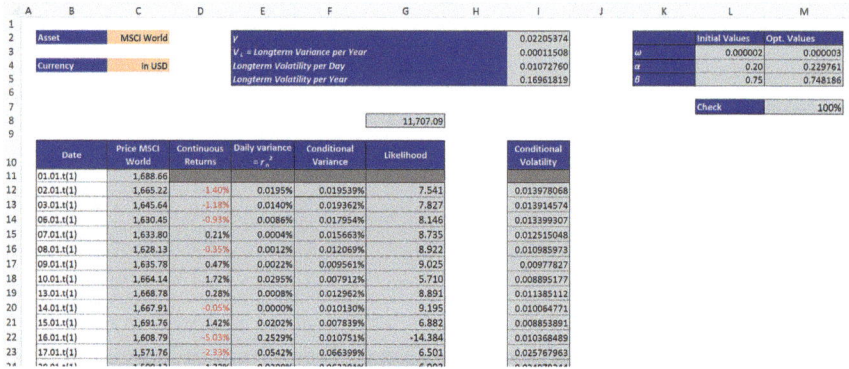

Fig. 2.23 Calculation of volatility with the GARCH model

Course Unit 2: Modelling Volatilities

Execution in Matlab

- Import the MSCI World prices and the associated dates.

```
Data = readmatrix('Matlab Data.xlsx');
MSCI = Data(:,3);
Date = Data(:,1);
Price_MSCI = [Date,MSCI];
Trading_days = 250;
```

- Calculate and plot the continuous returns of the MSCI World.

```
Continuous_Return = price2ret(Price_MSCI(:,2));
    T = length(Continuous_Return);
    figure
    plot(Continuous_Return)
    xlim([0, T])
    title('MSCI World returns')
    ylabel('Returns')
    xlabel('Days')
```

- Create a GARCH model with undefined parameters.

```
Model = garch(1,1);
```

- Fit the model to the continuous returns and estimate the parameters.

```
% Estimation of parameters using Max-Likelihood-Method
estMdl = estimate(Model,Continuous_Return(2:end),'E0',Continuous_Return(1));
Omega = estMdl.Constant % sprintf('%f',omega) for notation without e
Alpha = estMdl.ARCH{1}
Beta = estMdl.GARCH{1}
```

- Calculate the conditional variance as well as the conditional volatility, based on the determined parameters.

```
Daily_Variance = (Continuous_Return.^2)*100; % In percent -> following
values as well
Conditional_Variance = zeros(T,1);
Conditional_Variance(1) = Daily_Variance(1); % Initial values
for i = 2:T
Conditional_Variance(i)    =    Omega    +    Alpha*Daily_Variance(i-1)
+Beta*Conditional_Variance(i-1);
end
Conditional_Vola = sqrt(Conditional_Variance);
```

- Calculate the long-term variance and volatility per day and year.

```
Gamma = 1-Alpha-Beta; % Weighting factor average longterm volatility
    Variance_Day = Omega/Gamma; % Longterm variance per day
    Vola_Day = sqrt(Variance_Day); % Longterm volatility per day
    Vola_Year  =  Vola_Day*sqrt(Trading_days);  %  Longterm  volatility
per year
    Table        =        table(Gamma,Variance_Day,Vola_Day,
Vola_Year,'VariableNames',{'Gamma',      'Daily      Variance','Daily
Volatility','Annual Volatility'})
```

Matlab Results (Figs. 2.24 and 2.25)

```
GARCH(1,1) Conditional Variance Model (Gaussian Distribution):

                 Value        StandardError     TStatistic       PValue
                 _____      _____     _____       _____

Constant         2.3418e-06   5.9907e-07        3.909            9.2663e-05
GARCH{1}         0.76014      0.019702          38.582           0
ARCH{1}          0.22109      0.020151          10.972           5.2152e-28
```

Fig. 2.24 The estimated GARCH model

	Gamma	Daily Variance	Daily Volatility	Annual Volatility
1	0.0188	1.2475e-04	0.0112	0.1766

Fig. 2.25 Representation of the calculated long-term variance and volatility

Literature and Software References

> Hull, J. (2018). *Risk Management and Financial Institutions*. 5th ed. Wiley. pp. 227–229.
> McNeil., A., Frey, R., Embrechts, P. (2015). *Quantitative Risk Management: Concepts, Techniques and Tools*, Princeton University Press, pp. 112–118.
> See Excel file: Case Study Risk Management, Excel worksheet: GARCH.
> See Matlab script: A08_GARCH.

Course Unit 3: Modelling of Stochastic Processes

Assignment 9: Geometric Brownian Motion

Task

Model the MSCI WORLD index prices for the given data series for one year using the Geometric Brownian Motion.

Content

> GARCH models are deterministic processes, i.e. they depend on past values. This is particularly useful for very stationary assets, such as market interest rates. When forecasting longer-term price and value developments, stochastic processes lend themselves very well to taking the specifics and characteristics of historical price and value developments and using them for prediction purposes. A *stochastic process* (also called random process) is the mathematical description of temporally ordered, random processes. In financial market models, special stochastic processes are used which can be distinguished, for example, according to whether they are continuous or have jumps.
>
> For example, to simulate a stock index, we need models that allow *extrapolation* of price trends into the future and fit well with historical data and the general characteristics of price trends. We consider a forecast horizon of 1 year.
>
> From the multitude of stochastic processes, we select four processes that are of importance in risk management practice. These processes have different properties that make them more or less suitable for certain applications. According to these criteria, the appropriate process can be selected.
>
> The mean reversion property (reversion to the mean) states that asset volatility and historical returns eventually revert to the long-term mean or average level of the entire data set.
>
> Table 2.1 shows four stochastic process types with their properties.

Table 2.1 Process types and their properties

Name	Sign	Mean-Reversion	Long-term slope behaviour of the quantiles
Wiener Process	Positive and negative possible	No	Square root plus linear
Brownian Motion	Positive and negative possible	No	Square root
Geometric Brownian Motion	Only negative possible	No	Exponential
Vasicek Process	Positive and negative possible	Yes	Stationary

Important Formulas
Workbook: Case Study Risk Management Worksheet: GeoB

Wiener Process
A Wiener process is a continuous-time *stochastic process* that has *normally distributed* independent increments.

$$dW_t = \varepsilon \cdot \sqrt{dt} \qquad (2.23)$$

dW_t = Small change of the Wiener process
d = Differential
ε = Random variable that is standard normally distributed
t = Time course
dt = Small change of time

The Wiener process is characterised by the following properties:

- *Property 1*: $W_0 = 0$
- *Property 2*: The Wiener process $(W_t)_{t \geq 0}$ has independent increments, i.e. for the time $t_1 < t_2 < t_3 < \ldots < t_n$ the increments are $W_{t_n} - W_{t_{n-1}}$; $W_{t_{n-1}} - W_{t_{n-2}}$; $\ldots$; $W_{t_2} - W_{t_1}$ stochastically independent.
- *Property 3*: For all $0 \leq s < t$, it holds $W_t - W_s \sim N(0, t - s)$. The increments are normally distributed with expected value $E[W_t - W_s] = 0$ and variance $Var[W_t - W_s] = t - s$.
- *Property 4*: All (single) paths of $(W_t)_{t \geq 0}$ are continuous.

The Wiener process is a process with *Markov property*. The variable has "no memory", so that past realisations have no influence on the future development and are also not used to forecast future developments.

Brownian Motion
In addition to the Wiener process, Brownian Motion exhibits a drift μ and a volatility σ.

The performance in a Brownian Motion can be represented with the following formula:

$$S_t = \mu \cdot t + \sigma \cdot W_t \tag{2.24}$$

S_t = Random variable X at the time t
μ = Drift
t = Time course
σ = Volatility
W_t = Wiener process

Brownian Motion can also be represented as a differential equation:

$$dS_t = \mu \cdot dt + \sigma \cdot dW_t \tag{2.25}$$

dS_t = Small change of the random variable
dt = Small change of time
dW_t = Small change in the Wiener process

Brownian Motion was originally developed by the botanist Robert Brown to describe the motion of small particles in liquids or gases.

Geometric Brownian Motion

Geometric Brownian Motion, like Brownian Motion, is based on a Wiener process. It is:

$$S_t = S_0 \cdot \exp\left[\left(\mu_S - \frac{\sigma_S^2}{2}\right) \cdot t + \sigma_S \cdot dW_t\right] \tag{2.26}$$

S_t = Value at time t
S_0 = Start value at time 0
μ_S = Drift
σ_S = Volatility
W_t = Wiener process
t = Time

A typical application area is the modelling of stock prices. A helpful property for the modelling of stock prices is that the geometric Brownian Motion only takes on positive values. The reason for this is that the exponential function is always positive.

The geometric Brownian Motion can be described analogously with the following differential equation:

$$dS_t = S_t \cdot (\mu_S \cdot dt + \sigma_S \cdot dW_t) \quad (2.27)$$

S_t = Value at time t
dS_t = Small change of the value
d = Differential
dt = Small time interval

Here, dS_t describes a small change in the share price in a short time interval dt. The change results from the influence of two components: the drift rate μ_s and the volatility of the share σ_S. The drift rate μ_S is the average expected return of a stock. If $\mu_s > 0$, the value of the share price increases in expectation, if $\mu_s < 0$ the share price decreases in expectation. The volatility of the share σ_S controls the influence of chance, which is represented by the small change in the Wiener process, i.e. by dW_t. This differential Eq. (2.27) can be solved with the help of Itô's lemma (*Merton, Robert C. (1976) Option pricing when underlying stock returns are discontinuous*). The solution results in the representation in Eq. (2.26).

Geometric Brownian Motion is based on the following assumptions:

- $W_o = 0$
- The paths are continuous
- The increments of the Wiener process are stochastically independent and normally distributed, i.e. $W_t - W_s \sim N(0, \ t - s)$ for $0 \leq s \leq t$.

In the most popular financial mathematical models, Geometric Brownian Motion is used to model assets. Compare the Black-Scholes model for option pricing (Assignment 2.11) and the Merton model for calculating credit default probabilities (Assignment 2.17).

A few practical tips:

- Although the paths are continuous, they are modelled discretely for simplicity reasons.
- Since Brownian Motion is a future view, the volatility expected by the market is often used.

To model Brownian Motion in Matlab/Excel, we use the following simple representation:

$$S_t = S_0 L_1 L_2 \ldots L_n \quad \text{with } L_i = \frac{S(t_i)}{S(t_{i-1})} \quad (2.28)$$

Here, the elements are independent and lognormally distributed with expected value μ and variance σ^2.

This simple representation is valid because:

$$S_t = S_0 \cdot \frac{S(t_1)}{S(t_0)} \cdot \frac{S(t_2)}{S(t_1)} \cdot \ldots \cdot \frac{S(t_n)}{S(t_{n-1})}, \text{ with } S_0 = S(t_0) \text{ and } S_n = S(t_n) \quad (2.29)$$

It is:

$$S(t_i) = S_0\, e^{X(t_i)}, \text{ thus } L_i = \frac{s(t_i)}{s(t_{i-1})} = \frac{S_0\, e^{X(t_i)}}{S_0\, e^{X(t_{i-1})}} = e^{X(t_i) - X(t_{i-1})} \quad (2.30)$$

Since the increases $X(t_i) - X(t_{i-1})$ of a Brownian Motion are independent and normally distributed, the L_i are independent and lognormally distributed.

With the help of this representation, the creation of a geometric Brownian Motion can be easily implemented in Excel and Matlab.

Execution in Excel

- First, we transfer the variables "MSCI World Price", "Average Return" and "Implied Daily Volatility" (Fig. 2.26).
- Then one generates error terms in B10:B260 (for 250 trading days).
- Example: B10=NORM.INV(RAND(),$D$6,$D$7).
- In the cells C11:C260 you cumulate the values of column B.
- In cells D11:D260, the MSCI World price is multiplied by the exposed value in column C11:260.
- These MSCI World prices are then plotted as a line chart. The result is a possible simulation path.

Fig. 2.26 Assumptions of the Geometric Brownian Motion

Assumptions	
Price MSCI World in USD	3001.83
Mean Return	0.05%
Daily Historical Volatility	0.012

Excel Results (Fig. 2.27)

Fig. 2.27 Simulated prices of the MSCI World

Execution in Matlab
- Import the required data from the Excel file.

```
Data = readmatrix('Matlab Data.xlsx');
MSCI = Data(:,3);
Date = Data(:,1);
Price_MSCI = [Date,MSCI];
Data2 = readmatrix('Matlab Data.xlsx','Sheet','Black-Scholes');
Hist_daily_vola = Data2(3); % Historical daily volatility of the returns
```

Course Unit 3: Modelling of Stochastic Processes

- Determine the parameters required for the task.

```
Discrete_Return = price2ret(Price_MSCI(:,2),[],'Periodic');
Mean_Return = mean(Discrete_Return)
MSCI_current = MSCI(end) % Current price of the MSCI World index
Per = 240; % Number of periods
NumPath = 5; % Number of paths to be simulated
rng(1); % To replicate simulated values
T = ones(Per,1);
```

- Subsequently, simulate normally distributed random variables. These correspond to the variables L_1, L_2. ... Input for the simulation are the mean return, the volatility and the number of paths to be simulated. The values of the geometric Brownian Motion, at the respective time t, with the relation $S_t = S_0, L_1, L_2...L_n$ are then calculated. The result can be plotted with the Matlab function plot().

```
X = normrnd(Mean_Return,Hist_daily_vola,[Per, NumPath]);
MSCI_t = MSCI_current * exp(cumsum(X));
% MSCI(t+n) = MSCI (t)* (e^(Err(t))) * (e^(Err(t+1)) * (e^(Err(t+2)) ... (e^(Err(t+n))
plot(MSCI_t);
title('Simulated price of MSCI World');
ylabel('Price in USD');
xlabel('Time t');
```

Matlab Results (Fig. 2.28)

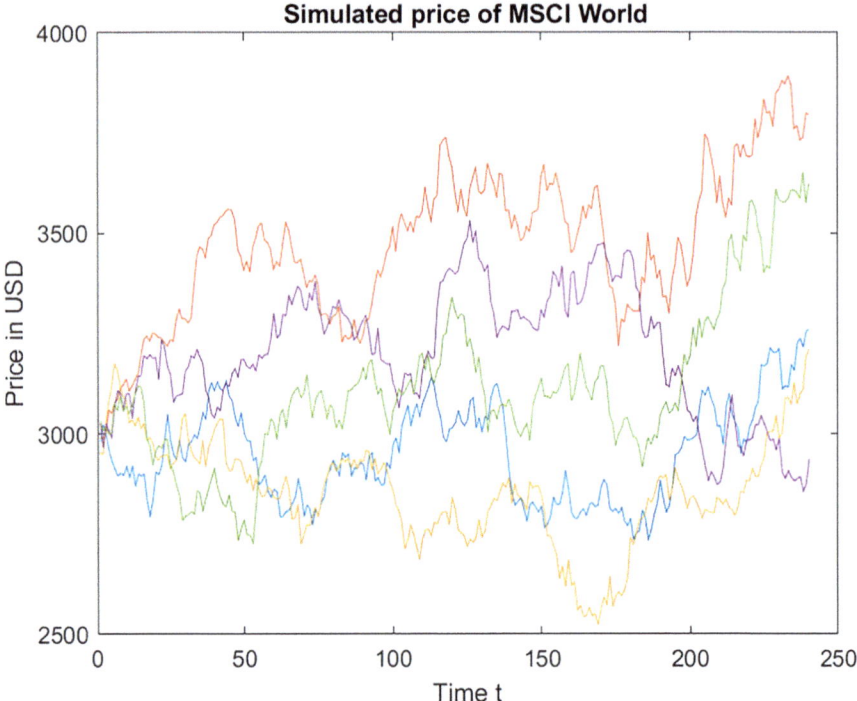

Fig. 2.28 Simulated paths of the MSCI World

Literature and Software References

Brzezniak, Z., Zastawniak, T. (2000). *Basic Stochastic Processes: A Course Through Exercises*, Springer, p. 151.

Gallager, R. G. (2013). *Stochastic Processes: Theory for Applications*. Cambridge University Press, pp. 142–144.

See Excel file: Case Study Risk Management, Excel worksheet: GeoB.

See Matlab script: A09_Geometric_Brownian_Motion.

Assignment 10: Vasicek/Ornstein-Uhlenbeck Process

Task

Model the EURIBOR Yields for the given data series for 3 months using the Vasicek/Ornstein-Uhlenbeck process.

Content

> Another class within the stochastic processes are the so-called *mean reversion processes*. Next to a diffusion parameter, there is a drift parameter which directs the process to a long-term mean reversion level (trend level). Mean reversion denotes that, for example, yields and interest rates are attracted by the mean value and return to the mean value in the long term.
>
> In the *Vasicek model* the instantaneous interest (short rate) is modelled using an *Ornstein-Uhlenbeck process*. Thus, the instantaneous interest has the mean reversion property, which means that the instantaneous interest is always attracted to the mean reversion level. If the random variable X is above the long term mean at a time t, then the drift term μ is negative, i.e. there is an attraction from above against the mean reversion level. If it is below, the drift term is correspondingly positive and there is an attraction from below against the mean reversion level. The parameter α determines the speed of the return to the mean reversion level. The parameter σ specifies the random influence of the Wiener process.

Important Formulas

Workbook: `Case Study Risk Management` Worksheet: `VasicekOrnstein`

The simplest example of a mean reversion process is the Ornstein-Uhlenbeck process:

$$dX_t = \alpha \cdot (\mu - X_t) \cdot dt + \sigma \cdot dW_t \qquad (2.31)$$

X_t	=	Random variable X at time t
d	=	Differential
α	=	Mean Reversion Speed = Stiffness
μ	=	Mean Reversion Level = Trend Level
σ	=	Diffusion = Volatility
dW_t	=	Small change of the Wiener process
t	=	Time
dt	=	Small time interval

If the diffusion is equal to 0, the random perturbation of the attraction at μ would be eliminated and the process would converge linearly towards the mean reversion level.

In the Vasicek model for interest rate modelling, the Ornstein-Uhlenbeck process is used. Therefore, it is also referred to as the Vasicek process.

In the following, we outline how parameters of the mean-reversion models can be estimated and subsequently simulated using linear regression.

A linear time series is known to be defined as follows:

$$y_t = a + b \cdot dt + \varepsilon_t \tag{2.32}$$

y_t = Value of the time series at time t
a = Constant
b = Slope
dt = Change of time
ε_t = Error term

The Vasicek process can be converted to the familiar linear form, which allows to use regression analysis to estimate the terms and obtain:

$$dX_t = \alpha \cdot \mu \cdot dt - \alpha \cdot X_t \cdot dt + \sigma \cdot dW_t \tag{2.33}$$

dX_t = y_t = Change of the assets
$\alpha \cdot \mu$ = a = Speed of the return
$\alpha \cdot X_t \cdot dt$ = $b \cdot dt$ = Amount of change
$\sigma \cdot dW_t$ = ε_t = "Random factor" which is to be minimised by the regression

Another well-known stochastic process similar in structure to the Vasicek process is the Black-Karasinski process. The implementation is similar, but the X_t is preceded by a logarithm.

Calculation of the probability distribution of ε_t:

Excel example: H18=NORM.DIST(F18,0,$D$8,FALSE)

Generation of W_t based on the optimised parameters:

Excel example: M17=NORM.INV(RAND(),0,$E$8)

Calculation of dX_t based on the simulated values:

Excel example: P18=$E$7*($E$6-Q17)+$E$8*O18

Course Unit 3: Modelling of Stochastic Processes 63

Determination of the simulated interest rates:

Excel example: Q17=D5+P17

Execution in Excel
- Determine dX_t in cells D18:D1320, as the difference of the current interest rate and the interest rate of the previous day D18=C18-C17.
- Calculate the term $\alpha(\mu - X_t)$ in cells E18:E1320, using the equation E18=$D$7*($D$6-C18).
- Subsequently, the error term ε_t can be calculated from $dX_t - \alpha(\mu - X_t)$. This is done in column F, F18=D18-E18 can be calculated. This is done in column F.
- To find the optimised parameter values, the probability distribution of the error term is determined in column H H18=NORM.DIST(F18;0;$D$8;FALSE).
- Subsequently, form the log-likelihood I18=LN(H18) and calculate the sum of these values in cell J18=SUM($I$18:$I$1320).
- Now use the Excel solver to calculate the optimised values of the parameters μ, α and σ. Select the sum of the log-likelihood as the target and maximise this value by changing the variable cells E6:E8.
- Based on the optimised values, interest rates of the 3M EURIBOR can then be simulated for a period of 100 days. For this purpose first W_t is generated as a random variable with an expected value of 0 and the optimised standard distribution M17=NORM.INV(RAND();0;$E$8). This is done in the cells M17:M116.
- Calculate in column N the differential of W_t N18=M18-M17 and cumulate this O18=O17+N18.
- In the cells P17:P116 you can then calculate dX_t using the formula above of the Ornstein-Uhlenbeck process P17=$E$7*($E$6-$D$5)+$E$8*$O$17 and P18=$E$7*($E$6-Q17)+$E$8*O18.
- Finally, future interest rates of the 3M EURIBOR are simulated Q17=D5+P17 and plotted as a line chart. The result is a possible simulation path.
- In addition, in cell D10=$D$5*EXP(-$D$7*D9)+$D$6*(1-EXP(-$D$7*D9)) the expected value of the 3M EURIBOR in 100 days can be determined.

Excel Results (Figs. 2.29, 2.30 and 2.31)

	A	B	C	D	E	F	G	H	I	J
14										
15		**Determination of Parameters based on Historical Data**						**Max-Likelihood**		
16		Date	Interest Rate	$dX_t = X_t - X_{t-1}$	$\alpha(\mu - X_t)$	$\varepsilon_t = X_{t+1} - X_t - \alpha(\mu - X_t)$		Probability Distribution	Log-Likelihood	Sum
17		01.01.t(1)	-0.263%							
18		02.01.t(1)	-0.264%	-0.00001	-2.93E-06	-7.07E-06		79.78838	4.37938	5706.26
19		03.01.t(1)	-0.262%	2E-05	-2.95E-06	2.29E-05		79.78762	4.37937	
20		06.01.t(1)	-0.263%	-0.00001	-2.94E-06	-7.06E-06		79.78838	4.37938	
21		07.01.t(1)	-0.262%	0.00001	-2.95E-06	1.29E-05		79.78819	4.37938	
22		08.01.t(1)	-0.264%	-2E-05	-2.93E-06	-1.71E-05		79.78799	4.37937	
23		09.01.t(1)	-0.265%	-0.00001	-2.93E-06	-7.08E-06		79.78838	4.37938	
24		10.01.t(1)	-0.266%	-0.00001	-2.92E-06	-7.08E-06		79.78838	4.37938	
25		13.01.t(1)	-0.266%	0	-2.92E-06	2.92E-06		79.78844	4.37938	
26		14.01.t(1)	-0.268%	-2E-05	-2.90E-06	-1.71E-05		79.78799	4.37937	
27		15.01.t(1)	-0.269%	-0.00001	-2.89E-06	-7.11E-06		79.78838	4.37938	
28		16.01.t(1)	-0.281%	-0.00012	-2.80E-06	-1.17E-04		79.76654	4.37910	
29		17.01.t(1)	-0.283%	-2E-05	-2.78E-06	-1.72E-05		79.78798	4.37937	
30		20.01.t(1)	-0.281%	2E-05	-2.80E-06	2.28E-05		79.78763	4.37937	
31		21.01.t(1)	-0.282%	-1E-05	-2.79E-06	-7.21E-06		79.78837	4.37938	
32		22.01.t(1)	-0.286%	-4E-05	-2.76E-06	-3.72E-05		79.78624	4.37935	
33		23.01.t(1)	-0.290%	-4E-05	-2.73E-06	-3.73E-05		79.78624	4.37935	
34		24.01.t(1)	-0.291%	-0.00001	-2.72E-06	-7.28E-06		79.78837	4.37938	
35		27.01.t(1)	-0.292%	-0.00001	-2.71E-06	-7.29E-06		79.78837	4.37938	
36		28.01.t(1)	-0.293%	-0.00001	-2.71E-06	-7.29E-06		79.78837	4.37938	
37		29.01.t(1)	-0.293%	0	-2.71E-06	2.71E-06		79.78844	4.37938	
38		30.01.t(1)	-0.293%	0	-2.71E-06	2.71E-06		79.78844	4.37938	
39		31.01.t(1)	-0.292%	0.00001	-2.71E-06	1.27E-05		79.78820	4.37938	

Fig. 2.29 Determination of the parameters of the Ornstein-Uhlenbeck process

	K	L	M	N	O	P	Q
14							
15			**Simulation of one Interest Rate Path**				
16		t	W_t	dW_t	Cumulative	dX_t	Simulated Interest Rates
17		1	0.00701	0.00701	0.00701	3.431E-05	-0.53957%
18		2	-0.00330	-0.01032	-0.00330	-1.729E-05	-0.54130%
19		3	0.00119	0.00449	0.00119	5.188E-06	-0.54078%
20		4	0.00508	0.00389	0.00508	2.464E-05	-0.53832%
21		5	0.00116	-0.00392	0.00116	5.009E-06	-0.53781%
22		6	-0.01183	-0.01299	-0.01183	-5.994E-05	-0.54381%
23		7	-0.00322	0.00861	-0.00322	-1.686E-05	-0.54549%
24		8	-0.00326	-0.00004	-0.00326	-1.706E-05	-0.54720%
25		9	0.00371	0.00697	0.00371	1.782E-05	-0.54542%
26		10	0.00486	0.00115	0.00486	2.354E-05	-0.54306%
27		11	-0.00442	-0.00927	-0.00442	-2.284E-05	-0.54535%
28		12	0.00251	0.00692	0.00251	1.180E-05	-0.54417%
29		13	-0.00750	-0.01001	-0.00750	-3.827E-05	-0.54799%
30		14	0.00019	0.00770	0.00019	2.539E-07	-0.54797%
31		15	-0.00188	-0.00208	-0.00188	-1.014E-05	-0.54898%
32		16	0.00664	0.00853	0.00664	3.251E-05	-0.54573%
33		17	-0.00074	-0.00738	-0.00074	-4.435E-06	-0.54617%
34		18	-0.00227	-0.00153	-0.00227	-1.210E-05	-0.54738%
35		19	0.01743	0.01970	0.01743	8.643E-05	-0.53874%
36		20	-0.00528	-0.02271	-0.00528	-2.719E-05	-0.54146%
37		21	0.00663	0.01191	0.00663	3.236E-05	-0.53822%
38		22	-0.00217	-0.00880	-0.00217	-1.165E-05	-0.53939%
39		23	0.00592	0.00809	0.00592	2.879E-05	-0.53651%

Fig. 2.30 Simulation of a possible path of future interest rates

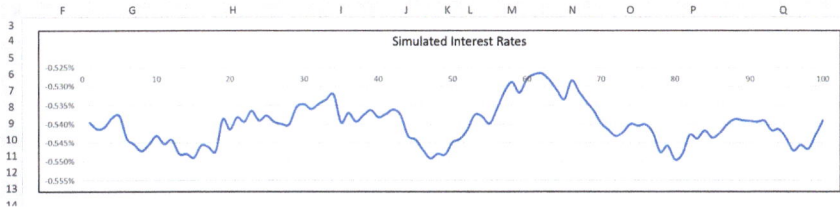

Fig. 2.31 Simulated interest rates for a period of 100 days

Execution in Matlab
- Import the required data from the Excel file.

```
Data = readmatrix('Matlab Data.xlsx');
ThreeM_EURIBOR = Data(:,2);
Date = Data(:,1);
EURIBOR = [Date,ThreeM_EURIBOR];
Return = ThreeM_EURIBOR;
```

- Define the number of periods and paths for the following simulation (discrete time steps are used for simplicity).

```
n = 100; % Number of periods
t = 20; % Number of paths to be simulated
rng(1); % To replicate simulated values
```

- Generate a matrix consisting of ones and the imported returns. The length of the matrix is determined by the number of returns -1 (degree of freedom).

```
Regressors = [ones(length(Return) −1, 1) Return(1:end-1)];
```

- Now the regressors can be simulated. To do so, use the function regress(Y,X), with the change of the returns diff() as the Y-value and the regressors as the X-value. Load the result into a matrix consisting of coefficients, intervals and residuals.

```
[Coefficients, Intervals, Residuals] = regress(diff(Return), Regressors);
```

- From this matrix the factors can be calculated. The change in time should be 1.

```
dt = 1; % Change in time =1
```

- The speed can be calculated by removing the factor dt analogous to $\alpha \cdot \mu \cdot dt$ (divide -Coefficients(2) by dt, but since d$t = 1$, this can be omitted).

```
Speed = −Coefficients(2); % Estimated speed
```

- Calculate the height by dividing −Coefficients(1) by −Coefficients(2).

```
Height = −Coefficients(1)/Coefficients(2); % Estimated height
```

- The sigma, which represents the error term, corresponds to the standard deviation of the residuals.

```
sigma = std(Residuals)/sqrt(dt); % Estimated standard deviation
```

- The calculated factors can now be loaded into an object. Choose a Hull-White-Vasicek hmv() object in this assignment.

```
obj = hwv(Speed, Height, sigma, 'StartState', Return(end))
```

- By loading the object into the Command Window, you will get an overview of the object and suitable simulation methods are suggested:

```
obj =

Class HWV: Hull-White/Vasicek
----------------------------------------
Dimensions: State = 1, Brownian = 1
----------------------------------------
StartTime: 0
StartState: -0.543
Correlation: 1
Drift: drift rate function F(t,X(t))
Diffusion: diffusion rate function G(t,X(t))
Simulation: simulation method/function simByEuler
```

```
Sigma: 0.00505004
Level: -0.642526
Speed: 0.000781885
```

- The object suggests the simulation method "simByEuler". Accordingly, simulate the object with this method and convert the resulting 3D matrix into a 2D matrix. Finally, represent the result graphically.

```
[X,T] = simByEuler(obj,n,'nTrials', t);
X =reshape(X,n+1,t);
plot(X)
title('Simulated 3M EURIBOR interest rate');
ylabel('Interest rate in %');
xlabel('Periods n');
xlim([0 100])
```

Matlab Results (Fig. 2.32)

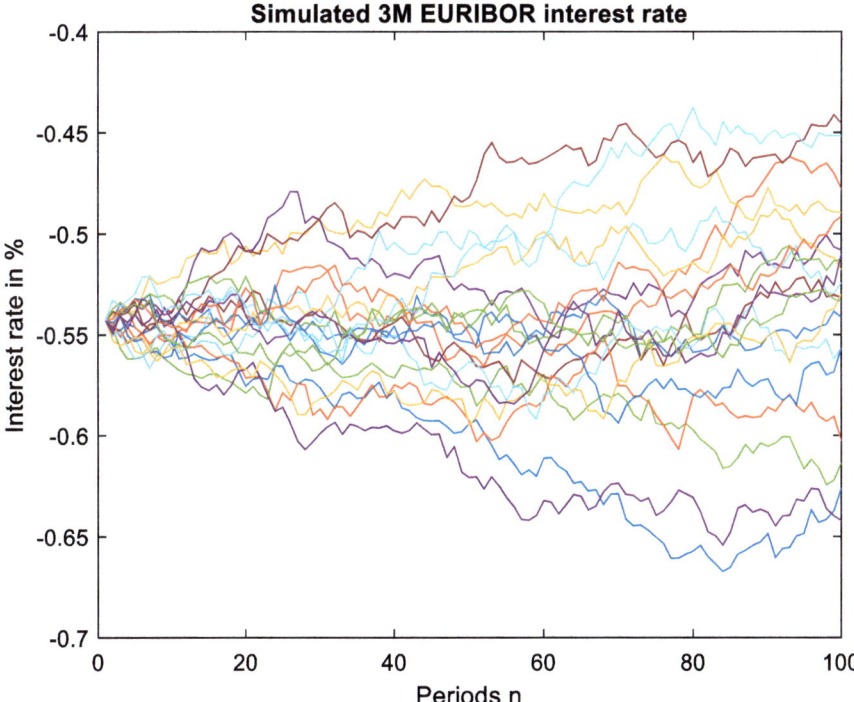

Fig. 2.32 Simulated 3M EURIBOR interest rates

Other models that can also be used as an object:

Brownian Motion (BM)
Geometric Brownian Motion (GBM)
Constant Elasticity of Variance (CEV)
Cox-Ingersoll-Ross (CIR)
Hull-White/Vasicek (HWV).
Heston
Merton
Bates

The following simulation methods can be selected:

- simulate = highly customisable simulation method
- simByEuler = Standard Euler Approximation
- interpolate = Brownian bridge (simulations with a fixed initial and final value very suitable for bonds).

Literature and Software References

> Brzezniak, Z., Zastawniak, T. (2020). *Basic Stochastic Processes: A course through Exercises, Springer, p. 220.*
> See Excel file: Case Study Risk Management, Excel worksheet: VasicekOrnstein.
> See Matlab script: A10_Vasicek_Ornstein.

Course Unit 4: Derivation of Risk Ratios with the Help of Black-Scholes

Assignment 11: From Geometric Brownian Motion to Black-Scholes

Task
Determine the fair price of a European call option with a term of 1 year using the Black-Scholes formula. The current price of the stock underlying the call option is 2,976 USD. In the historical data, a daily volatility of 1.2% was observed. The amount of 3,000 USD was contractually agreed as the strike price. The EURIBOR is currently 0%.

Course Unit 4: Derivation of Risk Ratios with the Help of Black-Scholes

Content

The *Black-Scholes model* is a financial mathematical model for *option price valuation*. The model was developed by the American economists *Fischer Black*, *Robert C. Merton* and *Myron S. Scholes*. Merton and Scholes were awarded the Nobel Prize in Economics in 1997 for their contribution to determining the value of an option. *Fischer Black* had already died by this time and therefore could not accept the prize.

An option is a contract that gives the buyer the right to purchase an underlying security at a fixed price (strike price) on a future date. The theory in the Black-Scholes model states that an investor can replicate an option through a portfolio of a stock and a risk-free security. The construction of such a portfolio is possible only if there is continuous trading in an efficient market without riskless arbitrage opportunities. This relationship can be modelled as a differential equation, also known as a heat equation from physics. The solution of this differential equation is the Black-Scholes formula for the valuation of European options.

The *input variables* in the Black-Scholes model are the current share price, the risk-free interest rate, the volatility, the strike price and the exercise date of the option. In order to apply the Black-Scholes model, the following assumptions are made:

– The stock price is modelled by a Geometric Brownian Motion with constant drift and volatility
– There are no arbitrage opportunities
– There is a continuous trade of the underlying
– There is a risk-free interest rate, which is constant for the term
– There are no transaction costs or taxes
– There are no dividend payments
– There are unrestricted short selling opportunities

The Black-Scholes model is a widely used financial mathematical model, although adjustments are often made to improve the model.

Important Formulas

Workbook: `Case Study Risk Management` Worksheet: `BlackScholes`

In the case of a call option, the buyer has the right, but not the obligation, to buy a share at a specific strike price and exercise date. In our example, the call option is far "out of the money", i.e. the market price of the stock is far below the strike price of the call option. The buyer can purchase the stock on the market at a lower price than the strike price. Therefore, it is not very likely that the call option will be exercised.

In the case of a put option, the buyer has the right, but not the obligation, to sell a share at a specific strike price and exercise date. The probability that the put option will be exercised in our case is thus higher than the probability that the call option will be exercised. Therefore, the put price is higher than the call price.

The price C_t of the call option at time t can be calculated under certain assumptions:

$$C_t = N(d_1)S_t - N(d_2)Ke^{-r(T-t)} \qquad (2.34)$$

S_t = Share price at current time t
K = Strike price
r = Risk-free interest rate
$T - t$ = Remaining term of the option with total term T at time t
N = Cumulative distribution function of the standard normal distribution

The price of the call option C_t at time t is described by two separate terms. The first part of the formula, $N(d_1)S_t$ is the expected value received when the stock is sold at the time the call option matures. The second part of the model, $N(d_2)Ke^{-r(T-t)}$, represents the present value of the payment of the strike price at the exercise date, assuming that the option is exercised. $N(d_2)$ can be interpreted as the probability that the option will be exercised. The fair market value of the call option thus depends on the difference between these two expressions.

The value d_1 results from the following equation:

$$d_1 = \frac{\ln\left(\frac{S_t}{K}\right) + \left(r + \frac{\sigma^2}{2}\right)(T-t)}{\sigma\sqrt{(T-t)}} \qquad (2.35)$$

σ = Volatility of the underlying asset

> Excel example: F4=(LN(C4/C5)+(C8+(C6^2)/2)*C7)/(C6*SQRT(C7))

The probability that the option will be exercised is calculated as follows:

$$d_2 = d_1 - \sigma\sqrt{T-t} \qquad (2.36)$$

Course Unit 4: Derivation of Risk Ratios with the Help of Black-Scholes 71

Excel example: F5=F4-C6*SQRT(C7)

Execution in Excel
- First, define the assumptions needed to calculate the option prices. The price of the underlying, the strike price, the historical daily volatility, the maturity in years and the EURIBOR interest rate (the risk-free interest rate) in cells C4:C8.
- Then calculate d_1 in cell F4 and d_2 in cell F5.
- In the next step, the cumulative value of the normal distribution at location d_1 is determined in cell F6 with NORM.DIST(F4,0,1,TRUE).
- The cumulative value of the normal distribution at the point d_2 is found in F7=NORM.DIST(F5,0,1,TRUE).
- Then, in cell F9, the corresponding values for $N(d_1)$ and $N(d_2)$ are substituted into the Black-Scholes formula =C4*F6-C5*(EXP(-C8*C7))*F7.

Excel Results (Fig. 2.33)

	A	B	C	D	E	F
1						
2			Black Scholes			
3						
4		Price Underlying	2,975.6 USD		d1	-2.3606
5		Strike Price	3,000.0 USD		d2	-2.3641
6		Historical daily Volatility	1.20%		Nd1	0.0091
7		Time to Maturity in Years	0.0833		Nd2	0.0090
8		EURIBOR interest rate	0.000%			
9					Call Price	0.0315 USD
10						
11					Put Price	24.4815 USD
12						
13					Put-Call Parity	24.4815 USD
14						
15						

Fig. 2.33 Calculation of the call price using the Black-Scholes formula

Execution in Matlab
- Import the required data and define the variables S0, K, r, t and Sig.

```
Data = readmatrix('Matlab Data.xlsx','Sheet','Black-Scholes');
    S0 = Data(1) % Price of underlying in USD
    K = Data(2) % Strikeprice in USD
    Sig = Data(3) % Historical daily volatility
    r = Data(4) % Risk-free interest rate
    t = Data(5) % Time to maturity in years
```

- Then enter the formulas for d1 and d2.

```
d1 = ((log(S0/K)+(r+((Sig^2)/2))*t)/(Sig*sqrt(t)));
d2 = d1-Sig*sqrt(t);
```

- Determine the price of the call option using the Black-Scholes formula.

```
CallPrice = S0.*normcdf(d1)-normcdf(d2).*K.*exp(-r*t)
```

- Alternatively, the formula from the Financial Toolbox can be used.

```
[Call,Put] = blsprice(S0,K,r,t,Sig)
```

Furthermore, Matlab allows to visualise the value/payout profile of a European Option:

- To do this, the price of the underlying "S0" in the previous code must be replaced by a vector. For example:

```
S0_range = [1:4000]';
d1_range = ((log(S0_range/K)+(r+((Sig^2)/2))*t)/(Sig*sqrt(t)));
d2_range = d1_range-Sig*sqrt(t);
```

- Since the asset prices are described by a vector of possible values, an element multiplication must be performed in Matlab. This requires the usage of the "." operator in the Matlab code.

```
CallPrice_range = S0_range.*normcdf(d1_range)-normcdf(d2_range).*K.
*exp(-r*t);
```

- This defines all prices of the call option depending on the value of the underlying. Now define the respective payouts at the corresponding points in time. For this, the option premium must be known. For simplification, assume a premium of 0.5 USD.

```
Optionpremium = 0.5;
```

- Accordingly, the payout "p" of the call option is defined as: $p = max(-0.5; S0 - K - premium)$. Translated into Matlab this results in:

```
PayoutCall = max(-(Optionpremium), S0_range-K-Optionpremium);
```

- Finally, plot the price and payout profile in comparison.

```
plot(CallPrice_range,'LineWidth',3)
hold on
plot(PayoutCall,'LineWidth',3)
xlim([2985 3015])
ylim([-1 16])
title('Payout profile Call Option');
ylabel('Payout/ Option price');
xlabel('Price underlying');
legend({'Call price' 'Payout'},'FontSize',7,'Location','NorthWest');
hold off
```

Matlab Results (Fig. 2.34)

Some properties of options can be derived from this representation. The terms an option is "in the money", "at the money" and "out of the money" are used to describe the value of an option. If the market price of the underlying stock is higher than the strike price of the call option, the buyer will exercise the call option. He thus makes a profit and the option is said to be "in the money". If the market price of the underlying stock is below the strike price of the call option, the buyer will not exercise the call option because he can buy the stock cheaper on the market. The option is then said to be "out of the money". An option is called "at the money" if the strike price is equal to the market price of the stock.

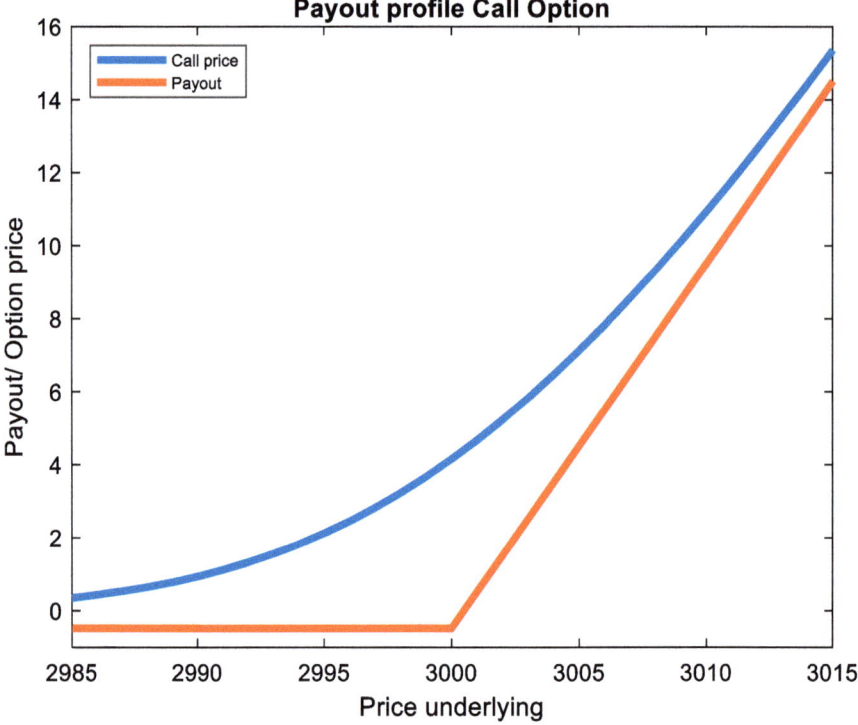

Fig. 2.34 Price of a call compared to its payout

Literature and Software References

- Hull, J. (2018). *Risk Management and Financial Institutions*. 5th ed. Wiley. pp. 673–675.
- See Excel file: Case Study Risk Management, Excel worksheet: BlackScholes.
  See Matlab script: A11_12_BlackScholes_PutCallParity.

Assignment 12: Excursus: Put-Call Parity

Task

Determine the fair price of a European put option with a term of one year using the Put-Call Parity. The underlying stock of the put option corresponds to the stock from Assignment 2.11. The EURIBOR is 0% as previously assumed.

Course Unit 4: Derivation of Risk Ratios with the Help of Black-Scholes

Content

> The *Put-Call Parity* describes the relationship between the price of a European call option and a European put option. Therefore, the price of a put option can be derived from the price of a call option. It should be noted, however, that both options must have the same maturity, as well as the same exercise price and the same strike price.
>
> The prices of the put option and the call option must satisfy the equation of the Put-Call Parity, otherwise there is a possibility of arbitrage.

Important Formulas

Workbook: `Case Study Risk Management` Worksheet: `BlackScholes`

The mathematical relationship between a put and a call is known as Put-Call Parity and is defined as:

$$C + K \cdot e^{-r \cdot T} = P + S_0 \tag{2.37}$$

S_0 = Share price at time 0
K = Exercise price (strike) of the option
r = Risk-free interest rate
T = Total term of the option
C = Price of the call option
P = Price of the put option

Thus, the put price P can be calculated by:

$$P = C + K \cdot e^{-r \cdot T} - S_0 \tag{2.38}$$

> Excel example: `F13=F9+(C5*EXP(-C8*C7))-C4`

Execution in Excel

- In cell `F11`, calculate the put price `F13=F9+(C5*EXP(-C8*C7))-C4`. This should be equal to the already determined put price based on the Black-Scholes formula.

Excel Results (Fig. 2.35)

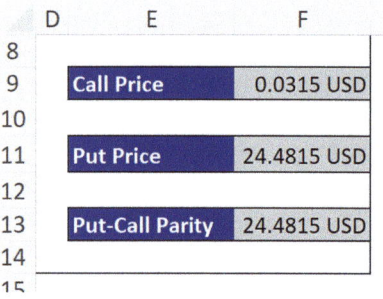

Fig. 2.35 Representation of the put price

Execution in Matlab
- Determine the price of a put option.

PutPrice = K.*exp(-r*t)*normcdf(-d2)-S0.*normcdf(-d1)

- The payout of a put is conversely to the payout of a call option.

PayoutPut = max(-(Optionpremium), K-S0_range-Optionpremium);

- Compute the value of a short future ($=P - C$), based on the Put-Call Parity.

Future = K*exp(-r*t)-S0_range;

- Display the results graphically.

```
plot(PayoutCall,'LineWidth',3)
hold on
plot(PayoutPut,'LineWidth',3)
hold on
plot(Future, '--','LineWidth',3)
xlim([2999 3001])
ylim([-1 1])
title('Put-Call Parity');
ylabel('Payout profile');
```

(continued)

```
xlabel('Price underlying');
legend({'Payout      Call'      'Payout      Put'      'Short
Future'},'FontSize',7,'Location','NorthWest');
   hold off
```

Matlab Results (Fig. 2.36)

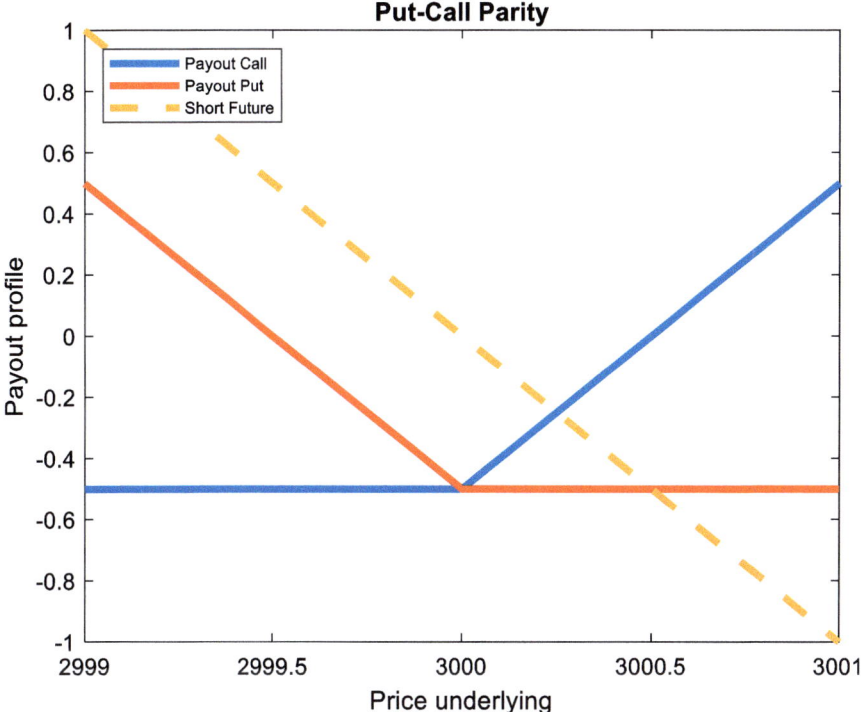

Fig. 2.36 Synthetic replication of a put option using a call and a short future

Literature and Software References

> Hull, J. (2018). *Risk Management and Financial Institutions*. 5th ed. Wiley. pp. 673–675.
> See Excel file: `Case Study Risk Management`, Excel worksheet: `BlackScholes`.
> See Matlab script: `A11_12_BlackScholes_PutCallParity`.

Assignment 13: Risk Metrics: The Greeks

Task

Calculate "the Greeks" of the call and put option from Assignment 2.11. Interpret the results.

Content

> The so-called *Greeks* are of great importance in risk management. They describe how strongly the price of an option depends on the input factors of the Black-Scholes formula. The most important Greeks are "Delta", "Gamma", "Vega", "Theta" and "Rho".
>
> The "*Delta*" risk figure indicates how much the price of the option changes if the strike price changes by one currency unit. Delta is therefore calculated using the first derivative of the Black-Scholes formula according to the strike price. The values of Delta for call options range between 0 and 1, and for put options between 0 and −1. A Delta of 0 means that there is no correlation between the option price and the strike price, whereas 1 or −1 means that the option price changes by one monetary unit if the strike price changes by one monetary unit.
>
> "*Gamma*" indicates how much the Delta changes when the base price changes by one currency unit. Gamma is therefore calculated via the second derivative of the Black-Scholes formula according to the strike price.
>
> "*Vega*" measures the impact of Implied Volatility on the option price. "*Theta*" describes how much the option price decreases over time. "*Rho*" indicates the sensitivity of the option price to a change in the risk-free interest rate.
>
> The "Greeks" thus provide information on price changes depending on changes in the strike price, underlying, Implied Volatility, risk-free interest rate and time.

Important Formulas

Workbook: `Case Study Risk Management` Worksheet: `BlackScholes`

Course Unit 4: Derivation of Risk Ratios with the Help of Black-Scholes

Delta
The Delta is the first derivative of the Black-Scholes formula according to the option's strike price. It indicates how much the price of an option changes when the strike price rises/falls by one monetary unit.

The following applies to the Delta of a call C with share price S:

$$\Delta_C = \frac{\partial C}{\partial S} = N(d_1) \geq 0 \tag{2.39}$$

and to the Delta of a put P:

$$\Delta_P = \frac{\partial P}{\partial S} = N(d_1) - 1 \geq 0 \tag{2.40}$$

S = Strike price
N = Distribution function of the standard normal distribution
d_1 = Definition of the value see Assignment 2.11

Gamma
The Gamma is the second derivative after the share price. It is the only "Greek of second order", which is counted among the classical Greeks.

It is valid for the Gamma of a call C and for the Gamma of a put P:

$$\Gamma_{C/P} = \frac{N'(d_1)}{S_0 \sigma \sqrt{T}} \geq 0 \tag{2.41}$$

S_0 = Share price at time 0
N' = Density function of the standard normal distribution
d_1 = Definition of the value see Assignment 2.11
σ = Volatility
T = Term of the option

Vega
Another factor influencing the Black-Scholes formula is volatility. Vega indicates the strength of the change in the option price with a change in the volatility of the underlying.

It is valid for the Vega of a call C and for the Vega of a put P:

$$V_{C/P} = S_0 \sqrt{T} N'(d_1) \geq 0 \tag{2.42}$$

S_0 = Share price at time 0
N' = Density function of the standard normal distribution
d_1 = Definition of the value see Assignment 2.11
T = Term of the option

Theta

Theta indicates the change in the option price over time. Since the probability of strong changes in the price of the underlying decreases with increasing time, Theta is always negative.

It is valid for the Theta of a call C:

$$\Theta_C = \frac{\partial C}{\partial t} = -\frac{S_0 N'(d_1)\sigma}{2\sqrt{T}} - rKe^{-rT}N(d_2) \tag{2.43}$$

and for the Theta of a put P:

$$\Theta_P = \frac{\partial P}{\partial t} = -\frac{S_0 N'(d_1)\sigma}{2\sqrt{T}} + rKe^{-rT}N(-d_2) \tag{2.44}$$

S_0 = Share price at time 0
N' = Density function of the standard normal distribution
d_1 = Definition of the value see Assignment 2.11
σ = Volatility
T = Term of the option
N = Distribution function of the standard normal distribution
d_2 = Definition of the value see Assignment 2.11
K = Strike price
r = Interest rate

Rho

Rho represents the last factor influencing the Black-Scholes formula, the market interest rate. It indicates the option price sensitivity to the market interest rate, i.e. shows how much the option price changes with rising/falling market interest rates.

It is valid for the Rho of a call C:

$$\rho_c = \frac{\partial C}{\partial r} = KTe^{-rT}N(d_2) \tag{2.45}$$

and for the Rho of a put P:

$$\rho_P = \frac{\partial P}{\partial r} = -KTe^{-rT}N(-d_2) \tag{2.46}$$

T = Term of the option
N = Density function of the standard normal distribution
d_2 = Definition of the value see Assignment 2.11
K = Strike price
r = Interest rate

Execution in Excel

- At the beginning, determine the individual terms that can subsequently be used for the calculation of all Greeks.
- Calculate $N(d_1)$ in cell I4=NORM.DIST(F4,0,1,TRUE) and $N(d_2)$ in cell I5=NORM.DIST(F5,0,1,TRUE).
- $N(-d_1)$, $N(-d_2)$ and $N'(d_1)$ are then to be determined using the formulas I6=NORM.DIST((-F4),0,1,TRUE), I7=NORM.DIST((-F5),0,1,TRUE) and I8=NORM.DIST((-F4),0,1,FALSE).
- Now calculate e^{-qT} in cell L4=EXP(-L9*C7), $S_0 e^{-qT}$ in cell L5=C4*L4 as well as e^{-rT} and Ke^{-rT} in cells L6=EXP(-C8*C7) and L7=C5*L6. q represents the dividend yield, which is assumed to be 0 in this assignment.
- Finally, determine $\sigma\sqrt{T}$ in cell L8=C6*SQRT(C7).
- Based on the individual terms, the Greeks for a call and put option can now be computed. The Delta of a call option is calculated using I12=I4*L4 and that of a put option L12=L4*(I4-1).
- Gamma, Vega and Theta do not differ for call and put options. Gamma is calculated in cell I13 and L13=I8/(C4*L8). Vega in cells I14 and L14=(C4*SQRT(C7)*I8) and Theta using =(-((C4*I8*C6)/(2*SQRT(C7)))-C8*L7*I5) in the corresponding cells I15 and L15.
- As the last Greek, determine the Rho of a call option in cell I16=C5*C7*L6*I5 and that of a put option in L16=-C5*C7*L6*I7.

Excel Results (Fig. 2.37)

	G	H	I	J	K	L
1						
2				Greeks		
3						
4		N(d1)		0.01	e^{-qT}	1.00
5		N(d2)		0.01	$S_0 e^{-qT}$	2975.55
6		N(-d1)		0.99	e^{-rT}	1.00
7		N(-d2)		0.99	$K*e^{-rT}$	3000.00
8		N'(d1)		0.02	σ√T	0.0035
9					Dividend in %	0.00
10						
11		Greeks (Call Option)			Greeks (Put Option)	
12		Delta		0.0091	Delta	-0.9909
13		Gamma		0.0024	Gamma	0.0024
14		Vega		21.127	Vega	21.127
15		Theta		-1.5211	Theta	-1.5211
16		Rho		2.2594	Rho	-247.7406
17						

Fig. 2.37 Calculation of the Greeks in Excel

Execution in Matlab
- Import the basic assumptions of the option.

```
Data = readmatrix('Matlab Data.xlsx','Sheet','Black-Scholes');
S0 = Data(:,1) % Price of the underlying in USD
K = Data(:,2) % Strikeprice in USD
Sig = Data(:,3) % Historical daily volatility
r = Data(:,4) % Risk-free interest rate
t = Data(:,5) % Time to maturity in years
```

- Calculate the Greeks of the options.

```
[CallDelta, PutDelta] = blsdelta(S0,K,r,t,Sig);
Gamma = blsgamma(S0,K,r,t,Sig);
Vega = blsvega(S0,K,r,t,Sig);
[CallTheta, PutTheta] = blstheta(S0,K,r,t,Sig);
[CallRho, PutRho] = blsrho(S0,K,r,t,Sig);
```

Matlab Results (Fig. 2.38)
- The Delta of the call option indicates that the price of the call option increases by 0.0091 USD if the value of the underlying increases by one USD. In contrast, the price of the put option decreases by 0.9909 USD.
- The sensitivity of the Delta of the options to a change in the value of the underlying is reflected in the Gamma. For every USD change in the value of the underlying, the Delta of the options will change by 0.0024.
- Vega represents the sensitivity of the option price to a change in volatility. If the volatility increases by one percent, the price of both options will change by 21.127 USD.
- The influence of the passage of time on the option price is given by Theta. Per past time unit, the price of both option types is reduced by 1.5211 USD.
- The option price sensitivity to the market interest rate is reflected in Rho. If the market interest rates rise by one percent, the price of the call option will rise by 2.2594 USD and the put option will fall by the corresponding amount.

		Delta	Gamma	Vega	Theta	Rho
1	Call Option	0.0091225	0.0023862	21.127	-1.5211	2.2594
2	Put Option	-0.99088	0.0023862	21.127	-1.5211	-247.74

Fig. 2.38 Calculation of the Greeks in Matlab

Course Unit 4: Derivation of Risk Ratios with the Help of Black-Scholes   83

Literature and Software References

- Hull, J. (2018). *Risk Management and Financial Institutions*. 5th ed. Wiley. pp. 161–162, 174–175.
- See Excel file: Case Study Risk Management, Excel worksheet: BlackScholes.
  See Matlab script: A13_Greeks.

Assignment 14: Implied Volatility—A Key Driver in Black-Scholes

Task
Determine the Implied Volatility of the MSCI World using the Black-Scholes formula.

Content

> In the previous assignment, a volatility has been assumed. In most cases, this volatility is based on historical data. This "historical volatility" is only suitable to a limited extent for the future. Therefore, the Implied Volatility is usually used for this purpose. *Implied Volatility* is the volatility assumed by the market. It is called Implied Volatility because it describes the volatility implied by option prices.
>
> The calculation of Implied Volatility is relatively simple. In practice, the maturity, the market interest rate, the strike price, the price of the underlying and the price of an exchange-traded call are known. Consequently, according to the Black-Scholes formula, only one variable is unknown: the volatility. If the Black-Scholes formula is rearranged according to the volatility, the result is the Implied Volatility.

Execution in Excel
Workbook: Case Study Risk Management Worksheet: Implied Volatility

- In cell C6, assume the historical volatility of 1.2% which has already been used in the previous assignment. This is used as the initial value and optimised in the next steps by a target value analysis.
- In order to calculate the Implied Daily Volatility assumed by the market, a "Goal Seek" is to be applied in Excel. You will find this in the "Data" tab, "Forecast" as a sub-item of the "What-if Analysis".

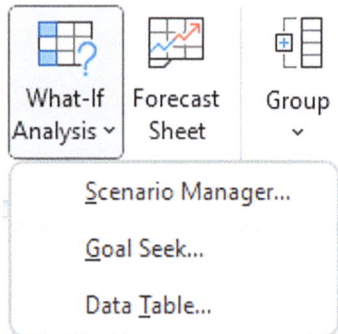

- The target value search determines the volatility assumed by the market, which when applied to the Black-Scholes formula yields the current price of a call option traded on the market.
- Enter as target cell F9, the determined price of the call option. This is to be brought to the target value of 2.01 USD by changing the volatility C6.

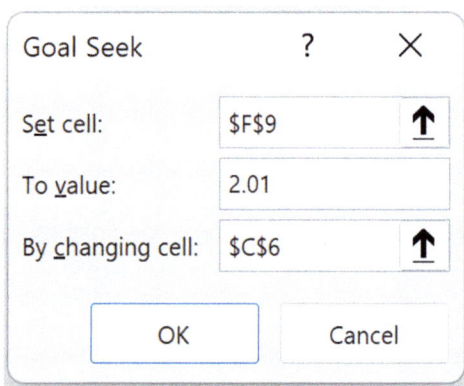

- Performing the target value search results in a calculated price of the call option of 2.01 USD, which coincides with the current market price. The Implied Volatility assumed for this is 2.82%.

Course Unit 4: Derivation of Risk Ratios with the Help of Black-Scholes

Excel Results (Fig. 2.39)

	A	B	C	D	E	F
1						
2		**Implied Volatility**				
3						
4		Price Underlying	2,975.55 USD		d1	-1.00
5		Strike Price	3,000.00 USD		d2	-1.01
6		Implied Daily Volatility	2.82%		Nd1	0.16
7		Time to Maturity in Years	0.0833		Nd2	0.16
8		EURIBOR interest rate	0.000%			
9		Price Call Option	2.01 USD		Call Price	2.01 USD
10						
11					Put Price	26.46 USD
12						

Fig. 2.39 The Implied Volatility and associated call prices

Execution in Matlab
- Import the basic assumptions of the option.

```
Data = readmatrix('Matlab Data.xlsx','Sheet','Black-Scholes');
S0 = Data(1) % Price of the underlying in USD
K = Data(2) % Strikeprice in USD
r = Data(4) % Risk-free interest rate
t = Data(5) % Time to maturity in years
CallPreis = Data(8) % Market price of a Call Option on the underlying above
```

- Determine the implied volatility with help of the function blsimpv() and the previously imported assumptions.

```
Implied_Vola = blsimpv(S0,K,r,t,CallPreis);
```

- This results in an Implied Volatility of 2.82%.

Literature and Software References

> Hull, J. (2018). *Risk Management and Financial Institutions*. 5th ed. Wiley. pp. 215–216.
>
> See Excel file: Case Study Risk Management, Excel worksheet: Implied Volatility.
> See Matlab script: A14_Implied_Vola.

Assignment 15: Volatility-Smile/-Surface

Task

Create a Volatility-Smile/Surface of the MSCI World.

Content

> The Implied Volatility calculation method leads to the Implied Volatility being constant across all maturities and strike prices. However, this rarely occurs in reality. Since the 1987 stock market crash, a phenomenon can be observed in equity options markets that was previously only known from FX options markets: the *Volatility-Smile*. The term comes from the fact that Implied Volatility as a function of the strike price produces a "U"-shaped curve, reminiscent of a smiling mouth.
>
> The explanations are manifold: behavioural economists argue that "out-of-the-money options" are a favourable hedging method for extremely optimistic/pessimistic investors, who thus react more strongly to swings (Shefrin 2008). Proponents of the market efficiency hypothesis reject this. *Merton*, for example, argues that the assumptions of the Black-Scholes model are too simplistic, especially the assumptions of the underlying Wiener process. If the logarithmic process is replaced by a jump model for strong swings, the Volatility-Smile could be explained (Merton 1976).
>
> Volatility-Smiles can also be extended by the dimension of maturities. Such a representation is called *Volatility-Surface*. It can be used to identify a rise or flattening of volatilities anticipated by the market. The Volatility-Smiles of different maturities are plotted one after the other and digitally combined to form a surface.

Course Unit 4: Derivation of Risk Ratios with the Help of Black-Scholes

Workbook: `Case Study Risk Management` Worksheet: `Volatility-Smile&-Surface`

Execution in Excel

- Refer to the Bloomberg data set in cells `B4:L26` (Fig. 2.40).
- To get the Volatility-Smile of the option with exercise date on 23.08.21 the corresponding cells `D6:L6` must be selected and the line chart has to be chosen under `Insert > Charts > Recommended Charts` and confirmed with OK. Afterwards, the label and the heading can be adjusted.
- To obtain the Volatility-Surface, first select the entire table area `D6:L26`.
- Then click `Insert > Charts > All Charts` and select the 3D surface.
- Excel shows the chart with the data series as the Z-axis. Usually, Volatility-Surfaces are displayed with the first smile in front. Therefore, the data must be inverted. To do this, right-click, on the `Chart Design > Select Data > Switch Row/Column`. Here you can also adjust the label of the horizontal axis with `Edit`. Then confirm with OK.

	A	B	C	D	E	F	G	H	I	J	K	L
1												
2					Implied Volatility/ Volatility-Smile and-Surface							
3												
4		Expiration Date	ImplTrm	80.0%	90.0%	95.0%	97.5%	100.0%	102.5%	105.0%	110.0%	120.0%
5				2380.1	2677.7	2826.4	2900.8	2975.2	3049.6	3123.9	3272.7	3570.2
6		23 Aug 2021	2965.83	33.87	25.07	20.35	18.32	16.66	15.41	14.64	14.68	19.12
7		20 Sep 2021	2964.6	36.74	25.17	20.42	18.41	16.71	15.33	14.31	13.45	16.46
8		18 Oct 2021	2963.03	34.97	24.59	20.37	18.58	17.03	15.75	14.75	13.71	15.88
9		20 Dec 2021	2960.63	31.47	23.47	20.2	18.79	17.55	16.5	15.65	14.63	15.61
10		21 Mar 2022	2954.66	28.35	22.3	19.81	18.71	17.71	16.83	16.06	14.93	14.52
11		20 Jun 2022	2944.76	27.42	22.5	20.33	19.32	18.38	17.52	16.74	15.47	14.47
12		19 Sep 2022	2938.21	26.37	21.97	20.09	19.24	18.44	17.7	17.02	15.86	14.59
13		19 Dec 2022	2932.39	25.6	21.55	19.89	19.15	18.45	17.8	17.2	16.1	14.42
14		20 Mar 2023	2926.92	25	21.22	19.75	19.11	18.5	17.94	17.41	16.42	14.68
15		19 Jun 2023	2917.87	24.5	20.96	19.64	19.06	18.52	18.02	17.54	16.64	14.91
16		18 Sep 2023	2914.99	24.02	20.72	19.57	19.07	18.61	18.18	17.77	16.98	15.21
17		18 Dec 2023	2913.11	23.85	20.71	19.59	19.09	18.63	18.2	17.79	17.01	15.33
18		24 Jun 2024	2903.93	22.6	20.15	19.42	19.12	18.84	18.58	18.32	17.75	16.03
19		23 Dec 2024	2903.79	22.91	20.38	19.56	19.21	18.88	18.58	18.29	17.7	16.2
20		22 Dec 2025	2901.49	21.95	20	19.56	19.39	19.26	19.13	19	18.68	17.34
21		31 Dec 2025	2901.8	21.95	20.02	19.58	19.42	19.28	19.15	19.03	18.71	17.38
22		31 Dec 2026	2904.28	23.37	21.91	21.42	21.2	21	20.81	20.62	20.23	19.23
23		31 Dec 2027	2910.71	24.83	23.65	23.1	22.84	22.58	22.33	22.09	21.64	20.81
24		31 Dec 2028	2920.03	24.95	23.93	23.45	23.22	23	22.79	22.58	22.18	21.47
25		31 Dec 2029	2931	25.04	24.14	23.72	23.52	23.32	23.14	22.95	22.6	21.97
26		31 Dec 2030	2943.57	25.12	24.31	23.93	23.75	23.58	23.41	23.24	22.93	22.36
27												

Fig. 2.40 Implied Volatility across multiple exercise prices at different exercise dates; data: Bloomberg received on: 20.07.2021

Excel Results (Figs. 2.41 and 2.42)

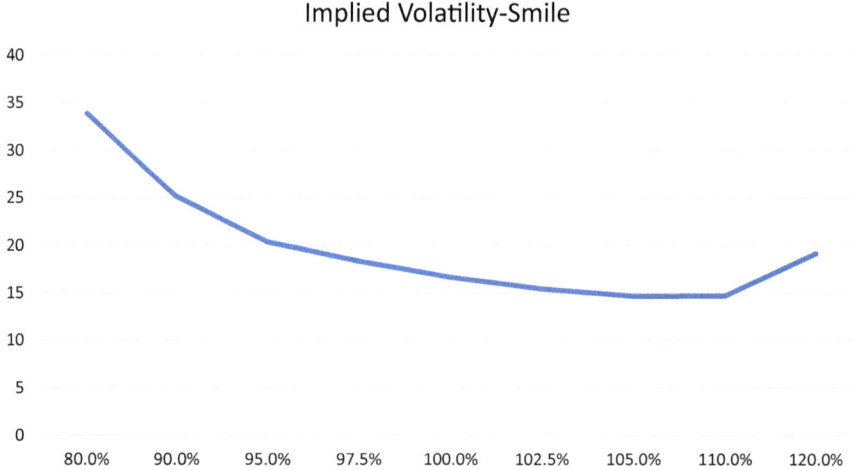

Fig. 2.41 Representation of an Implied Volatility-Smile

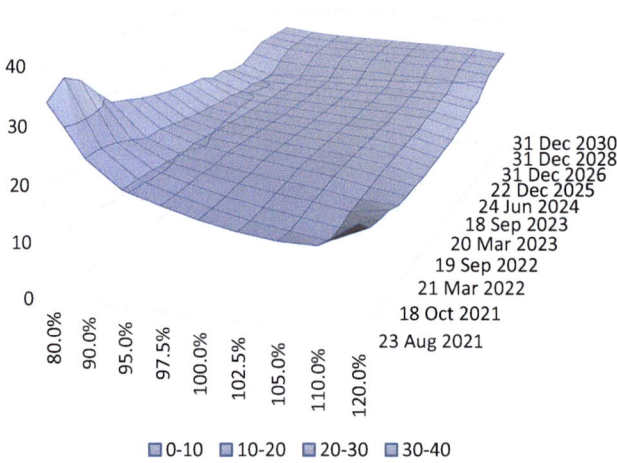

Fig. 2.42 Representation of the Implied Volatility-Surface

Execution in Matlab

Analogously, the Volatility-Smile can be implemented in Matlab.

- Load the same data set into the workspace area using drag & drop. The import window will then open. Select the data range (see Fig. 2.43) and choose "Numeric Matrix" as the output type. Confirm with "Import Selection".
- Alternatively, the dataset can be imported by code, similar to previous assignments.

Data = readmatrix('Matlab Data.xlsx','Sheet','Vola-Smile','RANGE','B4:J24');

#	A Verfallsdat...	B ImplTrm	C 80.0%	D 90.0%	E 95.0%	F 97.5% IVM1	G 100.0%	H 102.5%	I 105.0%	J 110.0%	K 120.0%
2			2.3801e+03	2.6777e+03	2.8264e+03	2.9008e+03	2.9752e+03	3.0496e+03	3.1239e+03	3.2727e+03	3.5702e+03
3	23 Aug 2021	2.9658e+03	33.8700	25.0700	20.3500	18.3200	16.6600	15.4100	14.6400	14.6800	19.1200
4	20 Sep 2021	2.9646e+03	36.7400	25.1700	20.4200	18.4100	16.7100	15.3300	14.3100	13.4500	16.4600
5	18 Oct 2021	2.9630e+03	34.9700	24.5900	20.3700	18.5800	17.0300	15.7500	14.7500	13.7100	15.8800
6	20 Dec 2021	2.9606e+03	31.4700	23.4700	20.2000	18.7900	17.5500	16.5000	15.6500	14.6300	15.6100
7	21 Mar 2022	2.9547e+03	28.3500	22.3000	19.8100	18.7100	17.7100	16.8300	16.0600	14.9300	14.5200
8	20 Jun 2022	2.9448e+03	27.4200	22.5000	20.3300	19.3200	18.3800	17.5200	16.7400	15.4700	14.4700
9	19 Sep 2022	2.9382e+03	26.3700	21.9700	20.0900	19.2400	18.4400	17.7000	17.0200	15.8600	14.5900
10	19 Dec 2022	2.9324e+03	25.6000	21.5500	19.8900	19.1500	18.4500	17.8000	17.2000	16.1000	14.4200
11	20 Mar 2023	2.9269e+03	25	21.2200	19.7500	19.1100	18.5000	17.9400	17.4100	16.4200	14.6800
12	19 Jun 2023	2.9179e+03	24.5000	20.9600	19.6400	19.0600	18.5200	18.0200	17.5400	16.6400	14.9100
13	18 Sep 2023	2.9150e+03	24.0200	20.7200	19.5700	19.0700	18.6100	18.1800	17.7700	16.9800	15.2100
14	18 Dec 2023	2.9131e+03	23.8500	20.7100	19.5900	19.0900	18.6300	18.2000	17.7900	17.0100	15.3300
15	24 Jun 2024	2.9039e+03	22.6000	20.1500	19.4200	19.1200	18.8400	18.5800	18.3200	17.7500	16.0300
16	23 Dec 2024	2.9038e+03	22.9100	20.3800	19.5600	19.2100	18.8800	18.5800	18.2900	17.7000	16.2000
17	22 Dec 2025	2.9015e+03	21.9500	20	19.5600	19.3900	19.2600	19.1300	19	18.6800	17.3400
18	31 Dec 2025	2.9018e+03	21.9500	20.0200	19.5800	19.4200	19.2800	19.1500	19.0300	18.7100	17.3800
19	31 Dec 2026	2.9043e+03	23.3700	21.9100	21.4200	21.2000	21	20.8100	20.6200	20.2300	19.2300
20	31 Dec 2027	2.9107e+03	24.8300	23.6500	23.1000	22.8400	22.5800	22.3300	22.0900	21.6400	20.8100
21	31 Dec 2028	2.9200e+03	24.9500	23.9300	23.4500	23.2200	23	22.7900	22.5800	22.1800	21.4700
22	31 Dec 2029	2931	25.0400	24.1400	23.7200	23.5200	23.3200	23.1400	22.9500	22.6000	21.9700
23	31 Dec 2030	2.9436e+03	25.1200	24.3100	23.9300	23.7500	23.5800	23.4100	23.2400	22.9300	22.3600

Fig. 2.43 Import Implied Volatilities into Matlab

- Select the first row of the matrix and plot it.

```
Smile = Data(1,:);
plot(Smile,'LineWidth',3)
title('Volatility-Smile');
ylabel('Option price');
xticklabels({'80.0%','90.0%','95.0%','97.5%','100.0%','102.5%',
'105.0%','110.0%','120.0%'})
xlabel('Volatility');
legend('Implied Volatility','FontSize',7,'Location','NorthEast');
```

- The Volatility-Surface can be graphically represented with the surf(X,Y,Z) function. For this, the data must first be prepared. Create the auxiliary vector T, which represents the number of periods and should be the same length as the data set, and an auxiliary vector D, with length of the first row of the data set.

```
T = 1:length(Data);
D = 1:length(Data(1,:));
```

- Subsequently, load the implied volatilities data into Z.

```
Z = Data;
```

- The coordinate system can be prepared with the meshgrid function.
- Finally, represent the surface as surf().

```
surf(X,Y,Z)
zlim([10 40])
xticklabels({'80.0%','90.0%','95.0%','97.5%','100.0%','102.5%',
'105.0%','110.0%','120.0%'})
    yticklabels({'23 Aug 2021','20 Jun 2022','18 Sep 2023','31 Dec 2025','31 Dec 2030'})
    title('Volatility-Surface');
    [X,Y] = meshgrid(D,T);
```

Course Unit 4: Derivation of Risk Ratios with the Help of Black-Scholes

Matlab Results (Figs. 2.44 and 2.45)

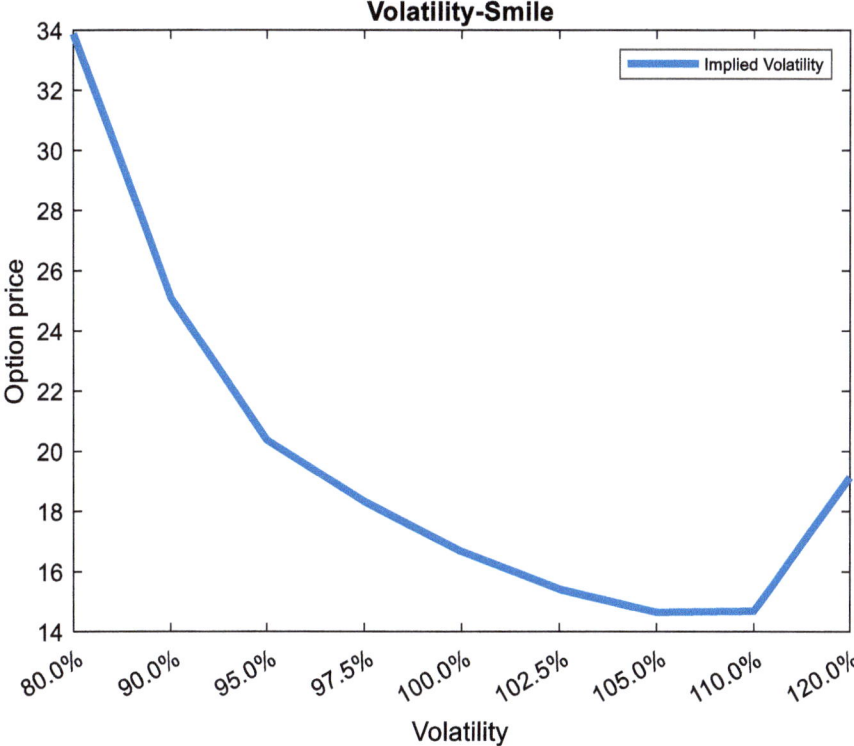

Fig. 2.44 Display of the Volatility-Smile in Matlab

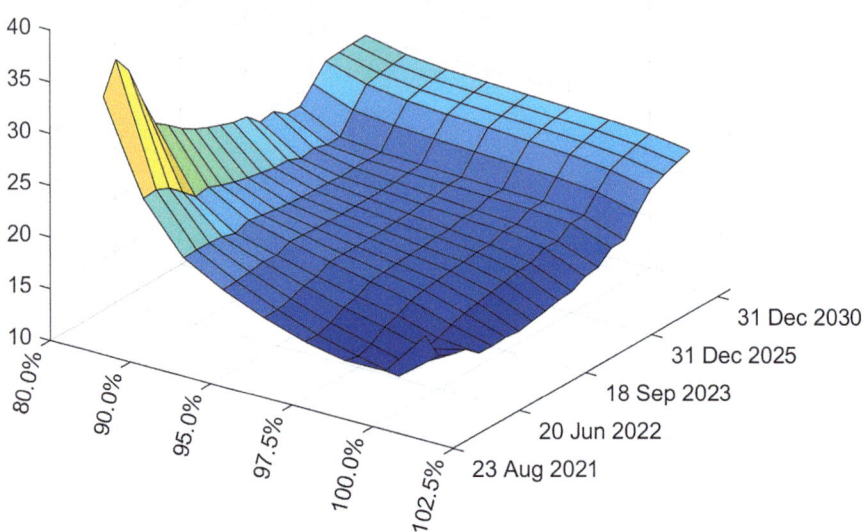

Fig. 2.45 Display of the Volatility-Surface in Matlab

Literature and Software References

> 📖 Hull, J. (2018). *Risk Management and Financial Institutions*. 5th ed. Wiley. pp. 575–577.
>
> ❌ See Excel file: Case Study Risk Management, Excel worksheet: Volatility-Smile&-Surface.
> See Matlab script: A15_Smile_Surface.

Chapter 3
Credit Risks

Credit risks describe possible losses resulting from a deterioration in the creditworthiness of a debtor, which includes in the worst case its default. One risk parameter for measuring credit risk is the *Probability of Default (PD)* within a certain period of time (often within 1 year). The convention is that a debt obligation is considered to be in default if the debtor is more than 90 days in arrears with the loan obligation and it is unlikely that the debtor will repay the debt.

There are various approaches for determining default probabilities. One approach is based on external and internal ratings (see Assignment 3.1), while other approaches are based on Merton's model, one of the best-known models in credit risk modelling (see Assignment 3.2).

Assignment 16: Rating Migration Matrices

Task

Determine the rating migration matrix after 2 years and after 5 years based on Moody's rating migration matrix for 1 year (Fig. 3.1).

Supplementary Information The online version contains supplementary material available at https://doi.org/10.1007/978-3-031-42836-4_3.

Initial Rating	Rating at Year End								
	Aaa	Aa	A	Baa	Ba	B	Caa	Ca-C	Default
Aaa	90.94%	8.36%	0.59%	0.08%	0.02%	0.00%	0.00%	0.00%	0.00%
Aa	0.87%	89.68%	8.84%	0.45%	0.07%	0.04%	0.02%	0.00%	0.02%
A	0.06%	2.64%	90.90%	5.67%	0.51%	0.12%	0.04%	0.01%	0.06%
Baa	0.04%	0.16%	4.44%	90.16%	4.09%	0.75%	0.17%	0.02%	0.18%
Ba	0.01%	0.05%	0.47%	6.66%	83.03%	7.90%	0.78%	0.12%	0.99%
B	0.01%	0.03%	0.16%	0.51%	5.32%	82.18%	7.39%	0.61%	3.79%
Caa	0.00%	0.01%	0.03%	0.11%	0.46%	7.82%	78.52%	3.30%	9.75%
Ca-C	0.00%	0.00%	0.07%	0.00%	0.80%	3.19%	11.41%	51.28%	33.24%
Default	0.00%	0.00%	0.00%	0.00%	0.00%	0.00%	0.00%	0.00%	100.00%

Fig. 3.1 Moody's rating migration matrix (1970–2016) via Hull, J. (2018)

Content

A popular way of representing the Probability of Default in practice is a *rating*. Rating agencies calculate a company's probability of downgrading using, among other things, key balance sheet figures, financial market data and expert opinions. The results are presented in a matrix, the so-called rating migration matrix.

In the *rating migration matrix*, the row shows the rating at the beginning of the year, while the column lists the rating at the end of the year. The rating migration matrix can be read as follows: A company that currently has an Aaa rating will be able to maintain its rating with a probability of 90.94%, but can also slip to Aa with a probability of 8.36%. Based on a rating migration matrix, however, future ratings can also be derived. For example, the probability that a company with an Aaa rating will still have an Aaa rating after 2 years. Here, the probability of a rating deterioration (red) must be multiplied by the probability of a readjustment to an Aaa rating (blue). This means that with a probability of $90.94\% \cdot 90.94\% = 82.7\%$ the rating will remain at Aaa after 2 years. With a probability of 8.36%, an Aaa rating changes to an Aa rating after 1 year. The probability of returning to an Aaa rating after another year is 0.87%. Overall, the probability of going from Aaa to Aa and back to Aaa is $8.36\% \cdot 0.87\% = 0.07\%$. Therefore, matrix multiplication can be used to calculate the rating after several years (Fig. 3.2).

Important Formulas

Workbook: Case Study Risk Management Worksheet: Rating Migration Matrix

From this calculation it can be deduced that the vector of row Aaa must be multiplied by the vector of column Aaa. To obtain the values for the further ratings, this procedure must be repeated for the respective rating. Consequently, the rating migration matrix M_n after n-years is an n-fold multiplication of the rating migration matrix after 1 year M_1.

Assignment 16: Rating Migration Matrices

Fig. 3.2 Rating migration matrix

Initial Rating	Rating at Year End								
	Aaa	Aa	A	Baa	Ba	B	Caa	Ca-C	Default
Aaa	90.94%	8.36%	0.59%	0.08%	0.02%	0.00%	0.00%	0.00%	0.00%
Aa	0.87%	89.68%	8.84%	0.45%	0.07%	0.04%	0.02%	0.00%	0.02%
A	0.06%	2.64%	90.90%	5.67%	0.51%	0.12%	0.04%	0.01%	0.06%
Baa	0.04%	0.16%	4.44%	90.16%	4.09%	0.75%	0.17%	0.02%	0.18%
Ba	0.01%	0.05%	0.47%	6.66%	83.03%	7.90%	0.78%	0.12%	0.99%
B	0.01%	0.03%	0.16%	0.51%	5.32%	82.18%	7.39%	0.61%	3.79%
Caa	0.00%	0.01%	0.03%	0.11%	0.46%	7.82%	78.52%	3.30%	9.75%
Ca-C	0.00%	0.00%	0.07%	0.00%	0.80%	3.19%	11.41%	51.28%	33.24%
Default	0.00%	0.00%	0.00%	0.00%	0.00%	0.00%	0.00%	0.00%	100.00%

$$M_n = M_1^n \tag{3.1}$$

n = Number of years
M_1 = Rating migration matrix after 1 year

Excel example: `C20=MMULT(C6:K14,C6:K14)`

Execution in Excel

- Refer to the rating migration matrix in cells `B4:K14`.
- Enter the matrix multiplication with the function `MMULT`: `C20=MMULT(C6:K14,C6:K14)` and confirm this with the key combination CTRL+Shift+Enter.
- An exponentiation of matrices cannot be implemented in Excel. Therefore, for longer durations, an extension of the `MMULT` function must be used. To display the rating migration matrix after 5 years in Excel, `C34=MMULT(C6:K14,MMULT(C6:K14,MMULT(C6:K14,MMULT(C6:K14,C6:K14))))` must be entered and confirmed with the key combination CTRL+Shift+Enter.

Excel Results (Figs. 3.3 and 3.4)

Initial Rating	Rating Migration Matrix after 2 Years								
	Rating at End of 2. Year								
	Aaa	Aa	A	Baa	Ba	B	Caa	Ca-C	Default
Aaa	82.77%	15.12%	1.82%	0.22%	0.05%	0.01%	0.00%	0.00%	0.00%
Aa	1.58%	80.73%	15.99%	1.32%	0.19%	0.09%	0.04%	0.00%	0.05%
A	0.13%	4.78%	83.12%	10.31%	1.13%	0.30%	0.09%	0.02%	0.14%
Baa	0.08%	0.41%	8.07%	81.82%	7.15%	1.63%	0.38%	0.04%	0.44%
Ba	0.02%	0.11%	1.13%	11.60%	69.64%	13.17%	1.87%	0.24%	2.24%
B	0.02%	0.06%	0.33%	1.25%	8.85%	68.56%	11.99%	1.06%	7.88%
Caa	0.00%	0.02%	0.07%	0.26%	1.19%	12.71%	62.61%	4.33%	18.80%
Ca-C	0.00%	0.00%	0.11%	0.09%	1.30%	5.21%	15.05%	26.69%	51.53%
Default	0.00%	0.00%	0.00%	0.00%	0.00%	0.00%	0.00%	0.00%	100.00%

Fig. 3.3 Moody's rating migration matrix after 2 years

Fig. 3.4 Moody's rating migration matrix after 5 years

Initial Rating	Rating at End of 5. Year								
	Aaa	Aa	A	Baa	Ba	B	Caa	Ca-C	Default
Aaa	62.75%	28.14%	7.58%	1.16%	0.20%	0.07%	0.02%	0.00%	0.03%
Aa	2.96%	60.27%	29.98%	5.27%	0.82%	0.33%	0.13%	0.01%	0.20%
A	0.40%	9.00%	65.74%	19.73%	3.19%	1.08%	0.32%	0.04%	0.55%
Baa	0.18%	1.45%	15.45%	63.40%	12.25%	4.33%	1.18%	0.13%	1.68%
Ba	0.06%	0.39%	3.57%	19.72%	43.71%	19.78%	5.10%	0.55%	7.16%
B	0.04%	0.16%	0.94%	3.83%	13.22%	43.21%	16.56%	1.65%	20.39%
Caa	0.01%	0.06%	0.26%	0.94%	3.46%	17.61%	34.25%	3.53%	39.88%
Ca-C	0.00%	0.03%	0.21%	0.51%	2.09%	7.30%	12.50%	4.53%	72.82%
Default	0.00%	0.00%	0.00%	0.00%	0.00%	0.00%	0.00%	0.00%	100.00%

Execution in Matlab

- First import the data of the rating migration matrix into Matlab. To do so, open the Excel file "Matlab Data" and drag and drop the rating migration matrix into the "Workspace" area.
- The import window opens automatically. To be able to use the data as a matrix, select the rows with numeric values and choose "Numeric Matrix" as "Output Type". Confirm with "Import Selection" (Fig. 3.5).

- Alternatively, the data set can be imported by code.

> Matrix = readmatrix('Matlab Data.xlsx','Sheet','Rating','RANGE','C6:K14')

- To obtain the rating migration matrix M2 after 2 years, multiply M by itself.

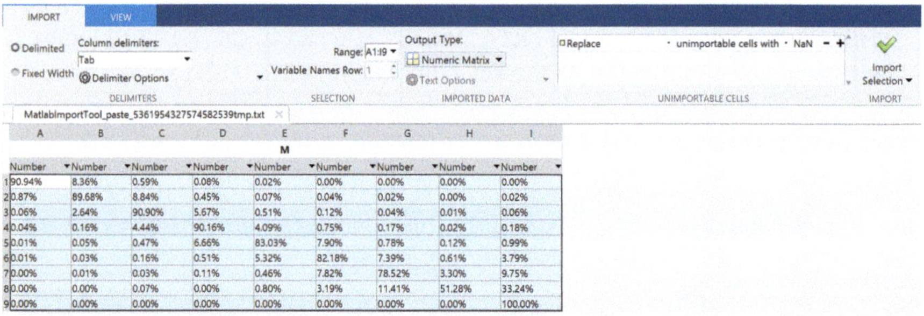

Fig. 3.5 Manual data import in Matlab

Assignment 16: Rating Migration Matrices

M2 = Matrix*Matrix

- Since Matlab is a matrix language, matrices can be exponentiated. To get the rating migration matrix M5 after 5 years, you need the 5th power of M.

M5 = Matrix^5

Matlab Results (Fig. 3.6)

From the rating migration matrix obtained, it can be seen that a company with an original Aaa rating will continue to have an Aaa rating after 5 years at 62.75% (cell (1,1)). Likewise, the probability that a company with an original rating of Ca-C will be insolvent after 5 years is 72.82% (cell (8,9)).

	1	2	3	4	5	6	7	8	9
1	0.6275	0.2814	0.0758	0.0116	0.0020	6.5486e-04	2.3111e-04	2.1048e-05	3.1224e-04
2	0.0296	0.6027	0.2998	0.0527	0.0082	0.0033	0.0013	1.3077e-04	0.0020
3	0.0040	0.0900	0.6574	0.1973	0.0319	0.0108	0.0032	4.3528e-04	0.0055
4	0.0018	0.0145	0.1545	0.6340	0.1225	0.0433	0.0118	0.0013	0.0168
5	6.1394e-04	0.0039	0.0357	0.1972	0.4371	0.1978	0.0510	0.0055	0.0716
6	3.8279e-04	0.0016	0.0094	0.0383	0.1322	0.4321	0.1656	0.0165	0.2039
7	6.9783e-05	5.5900e-04	0.0026	0.0094	0.0346	0.1761	0.3425	0.0353	0.3988
8	3.2593e-05	2.6068e-04	0.0021	0.0051	0.0209	0.0730	0.1250	0.0453	0.7282
9	0	0	0	0	0	0	0	0	1

Fig. 3.6 Calculated rating migration matrix after 5 years

Literature and Software References

Hull, J. (2018). *Risk Management and Financial Institutions*. 5th ed. Wiley. pp. 480–482.

McNeil., A., Frey, R., Embrechts, P. (2015). *Quantitative Risk Management. Concepts, Techniques and Tools*, Princeton University Press, pp. 374–375.

See Excel file: `Case Study Risk Management`, Excel worksheet: `Rating Migration Matrix`.

See Matlab script: `A16_Rating`.

Assignment 17: Merton's Model

Task

Calculate the Probability of Default of a company whose equity in market value is 3 million USD. The volatility of the equity is 80%. The company's liabilities of 10 million USD are due in exactly 1 year. Assume a risk-free interest rate of 5%.

Content

> *Merton's model* is one of the central risk management models in banking and insurance. It is used to calculate the Probability of Default of a company. In 1974, *Robert C. Merton* developed the idea of modelling the company's equity as a call option on its assets in order to assess the credit risk of a company. The performance of the assets is modelled by using a Geometric Brownian Motion. In addition, the company has liabilities. If, at the maturity date of the liabilities, the value of the assets is lower than the loans to be serviced, the company is bankrupt.

Important Formulas

As already discussed in Assignment 2.9, the change in the value of a company's assets S_t can be described with a Geometric Brownian Motion.

$$dS_t = S_t(\mu_S dt + \sigma_S dW_t), S_0 > 0 \qquad (3.2)$$

dS_t = Change in value of the company's assets in the period dt
dW_t = Change in the Wiener process in the period dt
μ_S = Average constant return of the company
σ_S = Average constant (implied) volatility of the company
S_o = Value of the company's assets at time 0

In Merton's model, it is assumed that the value of a company S_t at time t consists of the value of the equity capital E_t and that of the debt capital B_t.

$$S_t = E_t + B_t, 0 \leq t \leq T \qquad (3.3)$$

S_t = Value of the company's assets at time t
E_t = Value of equity at time t
B_t = Value of debt capital at time t

It is assumed that both E_t and B_t are frictionless, tradable securities. It is also assumed that the company does not pay any dividends or incur any new debt during the term.

Assignment 17: Merton's Model

The value of the debt capital B_t at time t is represented as a bond with a nominal value of K, which matures at time T. The nominal value K corresponds to the amount of the company's liabilities.

At repayment time T, the following scenarios may occur:

- $S_T > K$, i.e. at maturity T, the total value of the company's assets is greater than the debt represented by the nominal value K. In this case, the company can service the debtor. The shareholders receive the residual value. Thus, $B_T = K$ and $E_T = S_T - K$.
- $S_T < K$, i.e. at maturity T, the total value of the company's assets is less than the amount of its debts. Consequently, the company cannot service its debts and is insolvent at time T. The value of equity is therefore 0. Thus, $B_T = S_T$ and $E_T = 0$.

The mathematical representation of these equations is:

$$E_T = \max(S_T - K, 0) = (S_T - K)^+ \qquad (3.4)$$

$$B_T = \min(S_T, K) = K - (K - S_T)^+ \qquad (3.5)$$

The value of a company's equity can therefore be represented by a European call option on the value of the company, with the strike price equal to the amount of the liability.

The value of the debt is the liability less a European put option on the value of the company with a strike price equal to the liability.

Therefore, the Black-Scholes formula method can be used to determine the current price of the equity (equation I):

$$E_0 = S_0 \cdot N(d_1) - K e^{-rT} \cdot N(d_2) \qquad (3.6)$$

E_o = Value of equity at time 0
S_o = Value of the company's assets at time 0
$N(d_1)$ = Value of the standard normal distribution at d_1
K = Amount of liabilities due at maturity T
T = Maturity
$N(d_2)$ = Cumulative distribution function of the standard normal distribution at d_2

with

$$d_1 = \frac{\log\left(\frac{S_0}{K}\right) + \left(r + \frac{\sigma_S^2}{2}\right)T}{\sigma_S \sqrt{T}} \qquad (3.7)$$

and

$$d_2 = d_1 - \sigma_S \sqrt{T} \qquad (3.8)$$

The Probability of Default is given by $N(-d_2)$ if $\mu_S = r$, since:

$$P(S_t < K) = N\left(\frac{\log\left(\frac{K}{S_0}\right) - \left(r - \frac{\sigma_s^2}{2}\right)T}{\sigma_s\sqrt{T}}\right) = N(-d_2) \quad (3.9)$$

and

$$-d_2 = \frac{-\log\left(\frac{S_0}{K}\right) - \left(r + \frac{\sigma_s^2}{2}\right)T + \sigma_s^2 T}{\sigma_s\sqrt{T}} = \frac{\log\left(\frac{K}{S_0}\right) - \left(r - \frac{\sigma_s^2}{2}\right)T}{\sigma_s\sqrt{T}} \quad (3.10)$$

In accordance with the economic interpretation, the Probability of Default $N(-d_2)$:

- increases as the amount of payment obligations K increases,
- decreases with increasing volatility σ_s and increasing maturity T,
- decreases as the initial value of the company S_0 decreases, provided that the initial value is higher than the value of the liabilities, $S_0 > K$.

However, one last building block is still missing. Since in practice both the value of the company S_0 and the standard deviation of S_0, σ_S are unknown, they have to be calculated. To do this, we use Ito's lemma and obtain (equation II):

$$E_0 \sigma_E = \sigma_S N(d_1) S_0 \quad (3.11)$$

This approach yields two equations with two unknown variables.
Set Eq. (3.6) equal to zero as the function $F(S_0, \sigma_s)$ with:

$$F(S_0, \sigma_s) = S_0 N(d_1) - Ke^{-rT} N(d_2) - E_0 = 0 \quad (3.12)$$

Set Eq. (3.11) equal to zero as the function $G(S_0, \sigma_S)$ with:

$$G(S_0, \sigma_S) = \frac{\sigma_S}{\sigma_E} N(d_1) S_0 - E_0 = 0 \quad (3.13)$$

These two equations with the two unknowns S_0 and σ_s cannot be solved analytically. Therefore, we use the Least Squares Method as an approximation to solve this system of equations. For this, the following minimisation problem must be solved:

$$\min \; [F(S_0, \sigma_S)]^2 + [G(S_0, \sigma_S)]^2 \quad (3.14)$$

Execution in Matlab

Step 1: Definition of the "Merton" function

- In the first step, use the command "function" to define a Matlab function that is dependent on the vector "x0" and name it "Merton". Matlab is supposed to calculate the equation $F^2 + G^2$ using this function. In a later step, this function is to be minimised.

Assignment 17: Merton's Model

```
function [min] = Merton (x0);
global E0;
global SigmaE;
global r;
global T;
global K;
S0 = x0(:,1);
SigmaS = x0(:,2);
```

- The formulas for d_1, d_2, F and G are passed to Matlab, see the equations above.

```
d1 = (log(S0/K)+(r+((SigmaS^2)/2))/(SigmaS*sqrt(T)));
d2 = d1 - SigmaS*sqrt(T);
F = S0*normcdf(d1)-K*(exp(-r*T))*normcdf(d2)-E0;
G = (SigmaS/SigmaE)*normcdf(d1)*S0-E0;
```

- For the least squares method, the function $F^2 + G^2$ must be minimised. Here, however, only $F^2 + G^2$ is calculated. Subsequently, indicate the end of the function with the keyword "end".

```
min = (F^2)+(G^2);
end
```

- The function is saved under the name "Merton" in the same folder as the subsequent code.

Step 2: Transfer of the existing information to Matlab

- Open another script. First declare the input variables "E0", "SigmaE", "r", "T" and "K" as "global". Global variables can be used by several functions. Any change of value to a global variable is visible to all functions that declare it as global. Hence, the variables are known in the main script, as well as in the function "Merton".

```
global E0;
global SigmaE;
global r;
global T;
global K;
```

- Subsequently, the variables "E0", "SigmaE", "r", "T" and "K" are assigned numerical values. In this case, "E0" =3; "SigmaE" = 0.5; "r"=0.05; "T"=1 and "K"=10.

```
E0 = 3;
SigmaE = 0.5;
r = 0.05;
T = 1;
K = 10;
```

- The variables "S0" and "SigmaS" must also be assigned an initial value for the optimisation. Start with "S0" =13 and "SigmaS" =0.5.

```
S0 = 13;
SigmaS = 0.5;
```

Step 3: Calculation

- Finally, the function must be called with "fun=@Merton". Then the values of "S0" and "SigmaS" are passed to the vector "x0". The function can now be minimised with the command "fminsearch" and receives the input values of "S0" and "SigmaS" under fun.

```
fun = @Merton;
x0 = [S0 SigmaS];
fun = fminsearch(fun,x0);
```

- The values obtained can now be used to calculate the Probability of Default. To do this, read the values of "S0" and "SigmaS" from "fun".

```
S0 = fun(:,1);
SigmaS = fun(:,2);
```

- Finally, d2 is calculated based on the previously determined values, according to Eq. 3.8, and then used to calculate the Probability of Default.

```
d1 = (log(S0/K)+(r+((SigmaS^2)/2)*T)/(SigmaS*sqrt(T)));
d2 = d1 - SigmaS*sqrt(T);
Prob_of_default = normcdf((log(K/S0)-(r-((SigmaS^2)/2)*T))/SigmaS*sqrt(T));
```

Matlab Results (Fig. 3.7)

	d1	d2	Prob_of_default
1	0.64952	0.49249	0.032268

Fig. 3.7 The calculated probability of default

Literature and Software References

Hull, J. (2018). *Risk Management and Financial Institutions.* 5th ed. Wiley. pp. 452–453.

McNeil., A., Frey, R., Embrechts, P. (2015). *Quantitative Risk Management. Concepts, Techniques and Tools*, Princeton University Press, pp. 380–386.

See Matlab script: `A17_Merton`.

Assignment 18: Vasicek Model—Calculation of the Worst-Case Default Rate

Task

The Probability of Default (PD) of the individual loans is on average 1%. All loans in the portfolio have the same correlation to each other. Calculate the "Worst-Case Default Rate" at the 99% confidence level, assuming a strong correlation of 0.5 and a slightly weaker correlation of 0.15. Note that the correlation can take on values between -1 and 1.

Content

The *Worst-Case Default Rate WCDR(α)* is the share of defaulted loans within 1 year that will not be exceeded with probability α. The Worst-Case Default Rate thus answers the question of the default rate in a portfolio that will not be exceeded with a very high probability (here α). If, for example, the WCDR (99.9%) is 10%, it implies that: the bank can be 99.9% sure that the default rate in the portfolio will not exceed 10% within 1 year. If the correlation of the loans within a portfolio is zero, the Worst-Case Default Rate corresponds to the Probability of Default.

To calculate the Worst-Case Default Rate, it is assumed that all loans in a portfolio have the same 1-year Probability of Default. The literature shows that this is indeed approximately true for large portfolios.

The impact on the entire *loan portfolio* must be analysed if a company in the loan portfolio went bankrupt and can no longer service the loan. For this purpose, it is assumed that this loan portfolio consists of 1000 corporate loans. The economic effects often affect several companies at the same time or

(continued)

occure within a short period of time. Therefore, the correlations between the individual loans must be modelled.

The *Vasicek model* assumes that loans in a portfolio have the same correlation to each other. This is a gross simplification. Assuming that there are 1000 loans in a portfolio and the correlations between each two loans are chosen differently, the model would be much more complex. Therefore, a constant correlation between the default times of two companies is used and denoted by ϱ. The correlations between the times of loan defaults can be modelled using a *Gaussian copula*. More information on copulas can be found in Chap. 6.

The model for calculating the WCDR was developed by *Vasicek in* 1987 and is currently used to determine the minimum regulatory capital requirement for credit risk.

Important Formulas

Workbook: `Case Study Risk Management` Worksheet: `Vasicek`

The formula for the Worst-Case Default Rate, i.e., the proportion of defaulted loans within 1 year that, with probability α will not be exceeded, is:

$$WCDR(\alpha) = N\left(\frac{N^{-1}(PD) + \sqrt{\varrho}N^{-1}(\alpha)}{\sqrt{1-\varrho}}\right) \qquad (3.15)$$

N = Distribution function of the standard normal distribution
N^{-1} = Inverse of the standard normal distribution
PD = Probability of Default
ϱ = Correlation
α = Probability for not exceeding

Determination of the Worst-Case Default Rate for a high correlation:

Excel example: `C9=NORMSDIST((NORM.S.INV($C$6)+SQRT($C$4) *NORM.S.INV($C$7))/SQRT(1-$C$4))`

Determination of the Worst-Case Default Rate for a low correlation:

Excel example: `C10=NORMSDIST((NORM.S.INV($C$6)+SQRT($C$5)*NORM.S.INV($C$7))/SQRT(1-$C$5))`

Execution in Excel
- Calculate the Worst-Case Default Rate for a high correlation in cell `C9`, based on the linked assumptions in cells `C4:C7`.
- In the next step, determine the Worst-Case Default Rate for a low correlation in cell `C10`.

Assignment 18: Vasicek Model—Calculation of the Worst-Case Default Rate

Excel Results (Fig. 3.8)

	A	B	C
1			
2		**Calculation of Worst Case Default Rate**	
3			
4		Correlation high	0.50
5		Correlation low	0.15
6		Default Rate	0.01
7		Alpha	0.999
8			
9		WCDR (Correlation high)	42.085%
10		WCDR (Correlation low)	11.026%

Fig. 3.8 The calculated WCDR for a low as well as a high correlation

Execution in Matlab
- Determine the required parameters.

```
PD = 0.01; % Annual Probability of Default
Rho_low = 0.15; % Asset correlation
Rho_high = 0.5;
x = [0.0000000001: 0.0001: 0.2];
Alpha = 0.999
```

- Determine the probability distribution of the default rate.

```
pd = makedist('Normal'); % Probability distribution
pdf_low  =  sqrt((1-Rho_low)/Rho_low)*exp(0.5*((icdf(pd,x)).^2-((sqrt(1-Rho_low)*icdf(pd,x)-icdf(pd,PD))/sqrt(Rho_low)).^2)); % Density function for PD
pdf_high  =  sqrt((1-Rho_high)/Rho_high)*exp(0.5*((icdf(pd,x)).^2-((sqrt(1-Rho_high)*icdf(pd,x)-icdf(pd,PD))/sqrt(Rho_high)).^2)); % Density function for PD
WCDR_low  =  (cdf(pd,((icdf(pd,PD)+sqrt(Rho_low)*icdf(pd,Alpha))/sqrt(1-Rho_low))))
WCDR_high  =  (cdf(pd,((icdf(pd,PD)+sqrt(Rho_high)*icdf(pd,Alpha))/sqrt(1-Rho_high))))

Table1 = table(WCDR_low,WCDR_high)
```

- Plot the density of the portfolio default rate.

```
figure
h1 = line((x*100),pdf_low,'LineWidth',3,'Color',"b");
h2 = line((x*100),pdf_high,'LineWidth',3,'Color',"k");
axis([0 4 0 120])
title('Density portfolio default rate');
ylabel('Density');
xlabel('Default rate');
legend([h1 h2 ],{'Rho 0.15' 'Rho 0.5'},'FontSize',7,'Location','NorthEast');
```

Matlab Results (Figs. 3.9 and 3.10)

Fig. 3.9 The calculated WCDR

	WCDR_low	WCDR_high
1	0.11026	0.42085

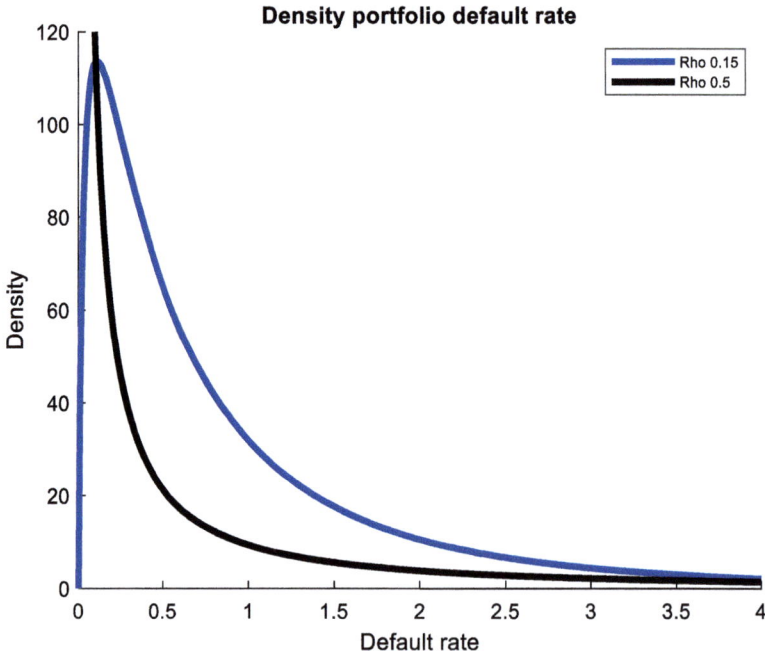

Fig. 3.10 Density of the portfolio default rate

Literature and Software References

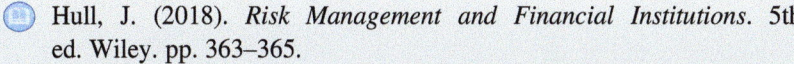

Hull, J. (2018). *Risk Management and Financial Institutions*. 5th ed. Wiley. pp. 363–365.

See Excel file: Case Study Risk Management, Excel worksheet: Vasicek.

See Matlab script: A18_19_Vasicek.

Assignment 19: Vasicek Model—Simulation of the Annual Portfolio Default Rate

Task

Simulate the annual portfolio default rate for a portfolio that can be assigned to retail business. Assume an annual Probability of Default (PD) of 1%.

Content

In the foundation-based approach of the internal models for determining the *regulatory minimum capital requirement* for credit risk, the banking supervisory authorities specify the formula for calculating the default correlation. The formula depends on the risk position class and is given as a function of the Probability of Default. For example, for retail business, the formula for default correlation is given below. The default correlation formulas are based on empirical research. The Probability of Default (PD) and the default correlation (ϱ) are inversely related to each other. If the Probability of Default of a company increases, the Probability of Default becomes more company-specific and is less influenced by general market conditions.

To simulate the *annual portfolio default rate*, its cumulative probability distribution must first be determined. The equation of the WCDR is resolved according to the probability α. A quick note for the statistics buffs, the WCDR is a quantile. By deriving the probability distribution, we obtain the probability density for the default rate.

Important Formulas

Workbook: Case Study Risk Management Worksheet: Vasicek

For the default correlation, use the formula specified in the basic approach of the internal models for determining the regulatory minimum capital requirement. This is the current one in the retail business:

$$\varrho = 0.03 \frac{1-e^{-35 \cdot PD}}{1-e^{-35}} + 0.16 \cdot \left(1 - \frac{1-e^{-35 \cdot PD}}{1-e^{-35}}\right) \qquad (3.16)$$

PD = Probability of Default

> Excel example: C17=0.03*((1-EXP(-35*$C$14))/(1-EXP(-35)))+0.16*(1-((1-EXP(-35*$C$14))/(1-EXP(-35))))

Depending on the risk position class, the default correlation is specified as a function of the Probability of Default.

Two transformations are necessary to simulate the annual portfolio default rate. In the first step, the cumulative probability distribution of WCDR must be determined. For this purpose, the equation of the WCDR is transformed according to α (probability that WCDR is less than or equal to x):

$$G(x) = N\left(\frac{\sqrt{1-\varrho}\, N^{-1}(x) - N^{-1}(PD)}{\sqrt{\varrho}}\right) \qquad (3.17)$$

$G(x)$ = Probability that the WCDR is less than or equal to x
N = Distribution function of the standard normal distribution
N^{-1} = Inverse of the standard normal distribution
PD = Probability of Default
ϱ = Default correlation

With a probability of $G(x)$, the WCDR is less than or equal to x. By deriving this function, we obtain the probability density for the failure rate:

$$g(x) = \sqrt{\frac{1-\varrho}{\varrho}} e^{\frac{1}{2}\left[\left(N^{-1}(x)\right)^2 - \left(\frac{\sqrt{1-\varrho}\, N^{-1}(x) - N^{-1}(PD)}{\sqrt{\varrho}}\right)^2\right]} \qquad (3.18)$$

> Excel example:
> C21=SQRT((1-$C$17)/$C$17)*EXP(0.5*NORMSINV(B21)^2-((SQRT(1-$C$17)*NORMSINV(B21)-NORMSINV($C$14))/SQRT($C$17))^2)

Determination of the Worst-Case Default Rate, using Eq. 3.15:

Assignment 19: Vasicek Model—Simulation of the Annual Portfolio Default Rate

Excel example:
C18=NORMSDIST((NORM.S.INV($C$14)+SQRT($C$17)*NORM.S.INV($C$15))/SQRT(1-$C$17))

Execution in Excel
- Calculate the default correlation in C17 based on the linked assumptions in cells C14:C15.
- In the next step, determine the Worst-Case Default Rate in cell C18.
- Calculate the probability density of the failure rate in cells C21:C155 for the given probabilities.
- Graph the calculated probability density.

Excel Results (Figs. 3.11 and 3.12)

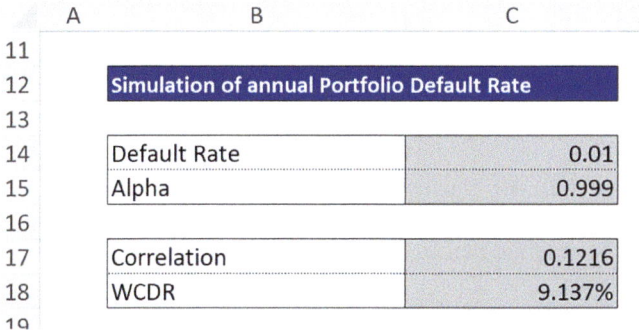

Fig. 3.11 Simulation of the annual portfolio default rate

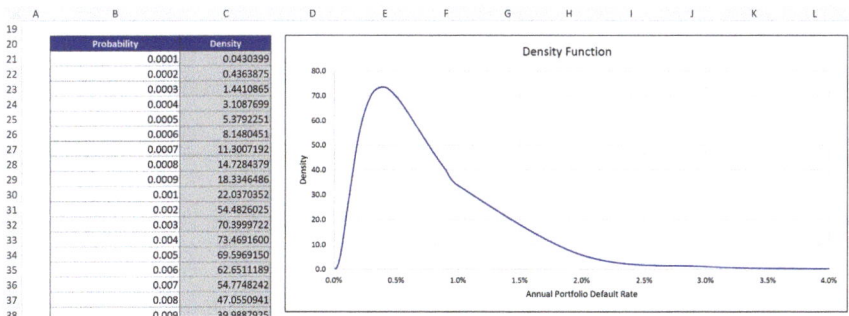

Fig. 3.12 Density of the annual portfolio default rate

Execution in Matlab
- Determine the annual portfolio default rate.

```
Rho_Mincapital = 0.03*((1-exp(-35*PD))/(1-exp(-35)))+0.16*(1-(1-exp(-35*PD))/(1-exp(-35)))
% Probability distribution
cdf_Mincapital = cdf(pd,((sqrt(1-Rho_Mincapital)*icdf(pd,x)-icdf(pd,PD))/sqrt(Rho_Mincapital)));
% Density function
pdf_Mincapital = sqrt((1-Rho_Mincapital)/Rho_Mincapital)*exp(0.5*((icdf(pd,x)).^2-((sqrt(1-Rho_Mincapital)*icdf(pd,x)-icdf(pd,PD))/sqrt(Rho_Mincapital)).^2));
WCDR_Mincapital = (cdf(pd,((icdf(pd,PD)+sqrt(Rho_Mincapital)*icdf(pd,Alpha))/sqrt(1-Rho_Mincapital)))) % In percent
Table2 = table(Rho_Min_Capital,WCDR_Minimum_Capital)
```

- Plot the density of the portfolio default rate.

```
figure
h3 = line((x*100),pdf_Mincapital,'LineWidth',3,'Color',"b");
axis([0 4 0 100])
title('Density portfolio default rate');
ylabel('Density');
xlabel('Default rate');
legend([h3],{'pdf'},'FontSize',7,'Location','NorthEast');
```

	Rho_Mincapital	WCDR_Mincapital
1	0.12161	0.091374

Fig. 3.13 The calculated default correlation and associated WCDR

Matlab Results (Figs. 3.13 and 3.14)

Assignment 19: Vasicek Model—Simulation of the Annual Portfolio Default Rate

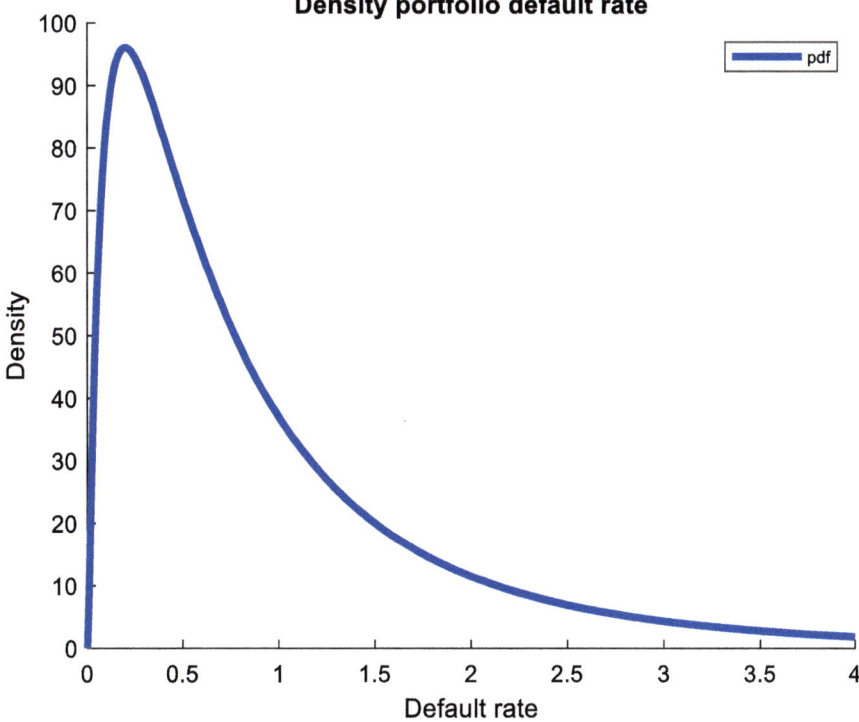

Fig. 3.14 Density of the portfolio default rate

Literature and Software References

> 📖 Hull, J. (2018). Risk Management and Financial Institutions. 5th ed. Wiley. pp. 260–263.
>
> ⓧ See Excel file: Case Study Risk Management, Excel worksheet: Vasicek.
> See Matlab script: A18_19_Vasicek.

Assignment 20: Vasicek Model—Estimation of Parameters from Historical Data

Task
Estimate the Probability of Default (PD) and the default correlation ϱ based on historical default rates. For this, refer to the data of the annual global study on corporate defaults and rating changes of the rating agency S&P. Assume a default correlation of 0.15 and a default probability of 1% as starting values.

Content

> In the previous assignments, the Probability of Default (*PD*) and the default correlation were chosen notionally. In the next step, these parameters are estimated based on historical data using the Maximum Likelihood Method. The *Maximum Likelihood Method* calculates the values of the parameters that maximise the probability of occurrence of the historical values (here default rates). The parameters *PD* and ϱ are determined in such a way that they best describe the observations that have occurred so far.
>
> The process is as follows:
>
> 1. Selection of initial values for the Probability of Default and the default correlation.
> 2. Computation of the logarithm of the probability density of the Probability of Default $g(x)$ based on historical data.
> 3. Calculation of the sum of these values.
> 4. Maximisation of this sum using the solver in Excel or the "fminsearch" command in Matlab to find the corresponding values of the probability of default (PD) and the default correlation.

Important Formulas
Workbook: Case Study Risk Management Worksheet: Vasicek
Determination of the log-likelihood:

$$g(x) = \ln\left(\sqrt{\frac{1-\varrho}{\varrho}} e^{\frac{1}{2}\left[\left(N^{-1}(x)\right)^2 - \left(\frac{\sqrt{1-\varrho}\, N^{-1}(x) - N^{-1}(PD)}{\sqrt{\varrho}} \right)^2 \right]} \right) \tag{3.19}$$

N^{-1} = Inverse of the standard normal distribution
PD = Probability of Default
ϱ = Default correlation

Assignment 20: Vasicek Model—Estimation of Parameters from Historical Data

Excel example: `D166=LOG(SQRT((1-$D$160)/$D$160)*EXP(0.5*NORMSINV(C166)^2-((SQRT(1-$D$160)*NORMSINV(C166)-NORMSINV($D$161))/SQRT($D$160))^2))`

Calculation of the sum of log-likelihoods:

Excel example: `D163=SUM($D$166:$D$205)`

Execution in Excel

- Link cells `B166:C205` to the annual data and associated annual historical default rates. These can be found in the worksheet "`Assumptions`" in cells `B280:C319`.
- Calculate the log-likelihood for the probability density for the probability of default $g(x)$ from the historical data, in cells `D166:D205`.
- Then calculate the sum of the log-likelihoods in cell `D163`.
- Link the cells of the optimised values `D160:D161` with the cells of the assumed initial values.
- Maximise the sum using the solver to find the default probability and default correlation values that maximise the sum.

Excel Results (Fig. 3.15)

	A	B	C	D
156				
157		Estimation of Parameters based on historical Data		
158				
159			Initial Values	Opt. Values
160		Correlation	0.150	0.119
161		Default Rate	0.010	0.017
162				
163				53.15
164				

Fig. 3.15 The estimated and optimised parameters, based on historical data

Execution in Matlab
- Import the required data.

```
Data = readmatrix('Matlab Data.xlsx','Sheet','Vasicek');
Default_rates = Data(:,2);
Years = 1980+[1:numel(Default_rates)].';
PD_annualy = [Years,Default_rates];
pd = makedist('Normal');
hat = [0.01 0.15];
PD_hat = hat(1)
Rho_hat = hat(2)
```

- Determine the required parameters, by applying the Maximum-Likelihood-Method. First, calculate the sum of logarithmic likelihoods.

```
LogLikelihood_hat = log10(sqrt((1-hat(2))/hat(2))*exp(0.5*norminv
(Default_rates).^2-((sqrt(1-hat(2))*norminv(Default_rates)-norminv(hat(1)))/
sqrt(hat(2))).^2));
Likelihood_hat = sum(LogLikelihood_hat)

Table1 = table(PD_hat,Rho_hat,Likelihood_hat)
```

- Maximise the sum of log-likelihood to find the optimal parameters.

```
fun = @(hat)sum(-log10(sqrt((1-hat(2))/hat(2))*exp(0.5*norminv
(Default_rates).^2-((sqrt(1-hat(2))*norminv(Default_rates)-norminv(hat(1)))/
sqrt(hat(2))).^2))); % Call Vasicek function
[Opt,SummeLogLikelihood,exitflag] = fminsearch(fun,[PD_hat Rho_hat]);
% Max-Likeli-Method
Likelihood_Opt = -(SummeLogLikelihood)
PD_opt = Opt(1)
Rho_opt = Opt(2)

Table2 = table(PD_opt,Rho_opt,Likelihood_Opt)
```

Assignment 21: Vasicek Model—Calculation of the Portfolio Loss

Matlab Results (Figs. 3.16 and 3.17)

	PD_hat	Rho_hat	Likelihood_hat
1	0.01	0.15	45.379

Fig. 3.16 Sum of log-likelihoods, based on the estimated parameters

	PD_opt	Rho_opt	Likelihood_Opt
1	0.017462	0.11915	53.151

Fig. 3.17 Sum of the log-likelihoods, based on the optimised parameters

Literature and Software References

Hull, J. (2018). *Risk Management and Financial Institutions*. 5th ed. Wiley. pp. 260–263.

See Excel file: Case Study Risk Management, Excel worksheet: Vasicek.

See Matlab script: A20_Vasicek.

Assignment 21: Vasicek Model—Calculation of the Portfolio Loss

Task

The assets consist of 100 million USD loans to A-rated companies. Calculate the portfolio loss (Value at Risk) that will not be exceeded with a probability of 99.9%. For this, use the assumptions and WCDR from Assignment 3.5. In addition, assume a loss given default (LGD) of 100% and the Exposure at Default (EAD) of 10 million USD.

Content

In addition to the information on the Worst-Case Default Rate calculated in Assignment 3.5, further information is required to calculate the portfolio loss. Firstly, the amount of receivables that is probably still outstanding at the time of default. This outstanding amount at default is referred to as *Exposure at Default* (*EAD*) and is an absolute monetary amount.

(continued)

On the other hand, the *loss ratio* is also included in the calculation. This is referred to as *Loss Given Default (LGD)*. The LGD is complementary to the recovery rate and is the proportion of receivables that will not be repaid in the event of default. The Loss Given Default (LGD) is estimated using historical, comparable loss data per rating grade. If banks use a standardised approach or the foundation approach of the internal models to determine the regulatory minimum capital requirement for credit risk, the banking supervisory authorities specify the values for Exposure at Default and Loss Given Default as default values.

The *portfolio loss (PL)* is the amount of money that will not be exceeded with a high probability (α). The portfolio loss is calculated from the Exposure at Default, the Loss Given Default and the Worst-Case Default Rate. The portfolio loss corresponds approximately to the Value at Risk.

Important Formulas
Workbook: Case Study Risk Management Worksheet: Vasicek

$$PL(\alpha) = EAD \cdot LGD \cdot WCDR(\alpha) \tag{3.20}$$

EAD	=	Exposure at Default
LGD	=	Loss Given Default
WCDR(α)	=	Worst-Case Default Rate with probability α

Determination of the Worst-Case Default Rate:

Excel example: C216=NORMSDIST((NORM.S.INV($C$211)+SQRT($C$210)*NORM.S.INV($C$212))/SQRT(1-$C$210))

Calculation of the portfolio loss (Value at Risk), which will not be exceeded with a probability of 99.9%:

Excel example: C217=$C$213*$C$214*$C$216

Execution in Excel
- Link the assumptions from the previous assignments and the assumed loss rate and Exposure at Default to cells C209:C214.
- Calculate the Worst-Case Default Rate in cell C216.
- Determine the portfolio loss (Value at Risk) that will not be exceeded with a probability of 99.9% in cell C217.

Assignment 21: Vasicek Model—Calculation of the Portfolio Loss

Excel Results (Fig. 3.18)

	A	B	C
206			
207		**Portfolio Loss**	
208			
209		Portfolio Volume	100,000,000 USD
210		Correlation	0.15
211		Default Rate	0.01
212		Alpha	0.999
213		Loss Given Default	1.00
214		Exposure At Default	10,000,000 USD
215			
216		WCDR	11.026%
217		Portfolio Loss	1,102,647.57 USD
218			

Fig. 3.18 The determined WCDR and the associated portfolio loss

Execution in Matlab

- Define the following assumptions.

```
PD = 0.01; % Annual Probability of Default
Rho = 0.15; % Asset correlation
Alpha = 0.999;
LGD = 1; % Loss Given Default
EAD = 10000000; % Exposure at Default
Portfolio = 100000000; % 100 Million USD
x = [0.0000000001: 0.0001: 0.2];
```

- Determine the Worst-Case Default Rate.

```
pd = makedist('Normal'); % Probability distribution
WCDR = cdf(pd,((icdf(pd,PD)+sqrt(Rho)*icdf(pd,Alpha))/sqrt(1-Rho)))
```

- Compute the portfolio loss.

```
PL = EAD*LGD*WCDR % Portfolio loss, that won't be exceeded with a
probability of 99.9%
```

Matlab Results (Fig. 3.19)

Fig. 3.19 The determined WCDR and the associated portfolio loss

	WCDR	PL
1	0.11	1102647.57

Literature and Software References

- Hull, J. (2018). *Risk Management and Financial Institutions*. 5th ed. Wiley. pp. 363–368.
- See Excel file: Case Study Risk Management, Excel worksheet: Vasicek.
  See Matlab script: A21_Vasicek.

Chapter 4
Operational Risks

Assignment 22: Calibration of the Loss Distribution Based on Expert Judgement

Task

Experts are asked about potential operational risks that could materialise for the company. After several rounds based on the Delphi method, the experts agreed on an average of 2 losses per year and estimated the most common value of the loss to be 1 million USD. A value above 10 million USD is considered extreme, however experts estimate that 5% of all events exceed this value.

Calibrate a distribution for the operational risks based on these expert judgements.

Content

> *"Operational risk"* is the risk of a loss resulting from inadequate or failed internal processes, people and systems or from external events. This definition includes legal risks, but does not include strategic risk or reputational risk (Basel Committee on Banking Supervision 2004).
>
> Operational losses can occur in *two forms*. On the one hand, there are operational risks that occur frequently but only lead to small losses. On the other hand, there are operational risks that occur extremely rarely but whose consequences can be so serious that they endanger the existence of a bank or insurance company. For the first type of operational risk, there is usually sufficient historical data available to model this risk. For the latter, the reverse

(continued)

Supplementary Information The online version contains supplementary material available at https://doi.org/10.1007/978-3-031-42836-4_4.

holds. Due to their rare occurrence, only a very limited amount of data is available.

If *timeseries of historical loss events* are available for individual operational risks, these can be used to calibrate a loss distribution, similar to market and credit risks. This approach alone is not optimal for operational risks. Especially for operational risks, appropriate controls are often implemented after a loss event has occurred. These are intended to prevent similar loss events from occurring in the future. Thus, the existence of historical loss events would be of little significance for the future.

Moreover, there is often also no sufficiently robust data history available to calibrate mathematical models. At this point, it is often useful and necessary to consult experts. An entire field of science is concerned with the question of how biases in the perception of risks can be minimised in order to improve the quality of the estimates. Numerous methods have been developed that consider psychological findings to reduce bias. One well-known method is the so-called *Delphi method*.

The aim of the *expert survey* is to identify relevant risks (referred to as scenarios) and to evaluate the individual scenarios in terms of loss frequency and the loss severity. Loss distributions are then calibrated on the basis of these estimates. In practice, the Poisson distribution is used to assess the loss frequency of operational risks. Only one parameter is needed, the average loss frequency within 1 year. This parameter can be determined with the following questions:

- In what period of time do you expect the damage to occur? 1 time in 1000 years, 1 time in 100 years, 1 time in 10 years....
- Or how many events do you expect to happen on average per year/in 10 years....
- How often, in what period of time can the damage occur?
- Answer: x times in y years

Previous research indicates that the first question often leads to an overestimation and question 2 to an underestimation of risk.

In addition to the loss frequency, the *loss amount distribution* is also relevant. In practice, the lognormal distribution is often used as the distribution function. To calibrate the lognormal distribution, it is recommended to ask for the "most frequently occurring" value. Furthermore, to ask for a value that is considered extreme and for the percentage of events that exceed this value.

Important Formulas

The loss frequency is modelled using the Poisson distribution. The Poisson distribution is suitable for modelling a number of events. The parameter λ of the Poisson distribution describes the expected frequency of the events. The probability that y losses occur per year is given by the following formula:

Assignment 22: Calibration of the Loss Distribution Based on Expert Judgement

$$p_\lambda(y) = \frac{\lambda^y e^{-\lambda}}{y!} \tag{4.1}$$

$p_\lambda(y)$ = Probability that there are y losses per year
λ = Expected number of losses per year

The loss severity distribution is modelled by the lognormal distribution. The lognormal distribution has two parameters μ and σ^2. The random variable X is lognormally distributed with the parameters μ and σ^2 if $ln(X)$ is normally distributed with the parameters μ and σ^2.

The experts answered the question about the most frequent value. This approximately describes the mode. For the mode M of a lognormal distribution the following formula holds:

$$M = e^{(\mu - \sigma^2)} \tag{4.2}$$

μ = Parameter of the lognormal distribution, mean of $ln(X)$
σ^2 = Parameter of the lognormal distribution, variance of $ln(X)$

The experts answered the question of a value that is considered to be extreme. This describes a quantile. In Chap. 5, we will refer to this as Value at Risk (VaR). A quantile is only meaningful if a confidence level α is specified. The confidence level α is answered by asking for the percentage of events exceeding the value considered to be extreme.

For the quantile/Value at Risk of the lognormal distribution at the confidence level α the following formula applies:

$$Q_\alpha = e^{(\mu + N^{-1}(\alpha) \cdot \sigma)} \tag{4.3}$$

$N^{-1}(\alpha)$ = Inverse of the standard normal distribution with confidence level α

In this way, a linear system of equations with two unknowns μ and σ^2 is obtained:

$$\begin{aligned}(I) & M = e^{(\mu - \sigma^2)} \\ (II) & Q_\alpha = e^{(\mu + N^{-1}(\alpha) \cdot \sigma)}\end{aligned} \tag{4.4}$$

To solve the linear system of equations, take the logarithm of both equations and subtract equation II from equation I. This results in:

$$0 = \sigma^2 + \sigma \cdot N^{-1}(\alpha) + ln(M) - ln(Q_\alpha) \tag{4.5}$$

This is a quadratic equation in σ and can be solved with the quadratic formula. Once σ is calculated, equation I gives the parameter μ.

Based on the determined parameters, the loss severity distribution can be estimated. It should be noted that a negative standard deviation σ is to be excluded. If the quadratic formula results in two possible σ, the loss distribution must be calculated for each σ. Unless one distribution can be excluded by simple exclusion criteria, the next step is to talk to experts again.

Execution in Matlab
- Define the assumptions obtained through expert interviews.

```
lambda = 2; % Average damages per year
M = 1000000; % Expected value of damage (1 Mio.)
Q = 3000000; % Extreme damage (3 Mio.)
alpha = 0.95; % Confidence interval
rng default
```

- Model the frequency of damages based on a Poisson distribution and graph the results.

```
% Modelling by Poisson distribution
x = 0:100;
Freq = poisspdf(x,lambda); % Damage frequency
bar(x,Freq)
xlim([-0.5,15])
title('Modelled frequency of damage per year')
xlabel('Frequency')
ylabel('Probability of occurrence')
```

- Write the ABC-formula into a manually created function. Either save it in the same folder or append the function to the end of the script.

```
function [sigma1,sigma2] = ABC_formula(a,b,c)
D = b^2-4*a*c; % Discriminant >0 -> 2 solutions; =0 -> 1 sol.; <0 -> no solution
if D>0
sigma1 = (-b+sqrt(D))/(2*a); % Solution x1
sigma2 = (-b-sqrt(D))/(2*a); % Solution x2
end
if D == 0
sigma1 = (-b+sqrt(D))/(2*a); % Solution x1
sigma2 = 0; % No second solution
end
end
```

Assignment 22: Calibration of the Loss Distribution Based on Expert Judgement

- Model the distribution of damage values. First determine σ and μ from equation 4.5, using the ABC-formula defined above.

```
% Modelling by lognormal distribution
% Solving linear system of equations by ABC-formula
% Coefficients
a = 1; % Factor before x^2
b = norminv(alpha); % Factor before x
c = log(M)-log(Q); % Remaining summands
[sigma1,sigma2] = ABC_formula(a,b,c); % Solution of ABC-formula (x1,x2)
% Alternative method "roots([a b c])" p-q-formula
sigma = [sigma1,sigma2]
mu = log(M)+sigma.^2
```

- Then determine the loss distributions for the two possible outcomes and calculate the corresponding expected value and quantile.

```
% Compound Poisson process
n = 1000; % Number of simulated damages
pois = @(m)poissrnd(lambda,m,1); % Poisson sample, with length m
logs = @(m)exp(mu(1)+sigma(1)*randn(m,1)); % Log-normal sample, with length m (x1)
logs2 = @(m)exp(mu(2)+sigma(2)*randn(m,1)); % Log-normal sample, with length m (x2)
y = zeros(n,1); % For x1
y2 = zeros(n,1); % For x2
for i = 1:length(y)
nr = pois(1);
for j = 1:nr
y(i) = y(i)+logs(1);
y2(i) = y2(i)+logs2(1);
end
end
M_x1 = mean(y); % Expected loss through operational risks per year
Q_x1 = quantile(y,alpha);
M_x2 = mean(y2); % For x2
Q2_x2 = quantile(y2,alpha);
format bank
comparison   =   table(M_x1,Q_x1,'VariableNames',{'Expected   loss'
'Quantile'},'RowNames',{'x1'})
```

- Finally, display the results and conduct additional expert interviews to determine the applicable loss distribution.

```
histogram(y)
title('Operational damages for x1')
xlim([0,15*10e5])
ylabel('Frequency')
xlabel('Damages in USD')
```

Matlab Results (Figs. 4.1, 4.2, and 4.3)

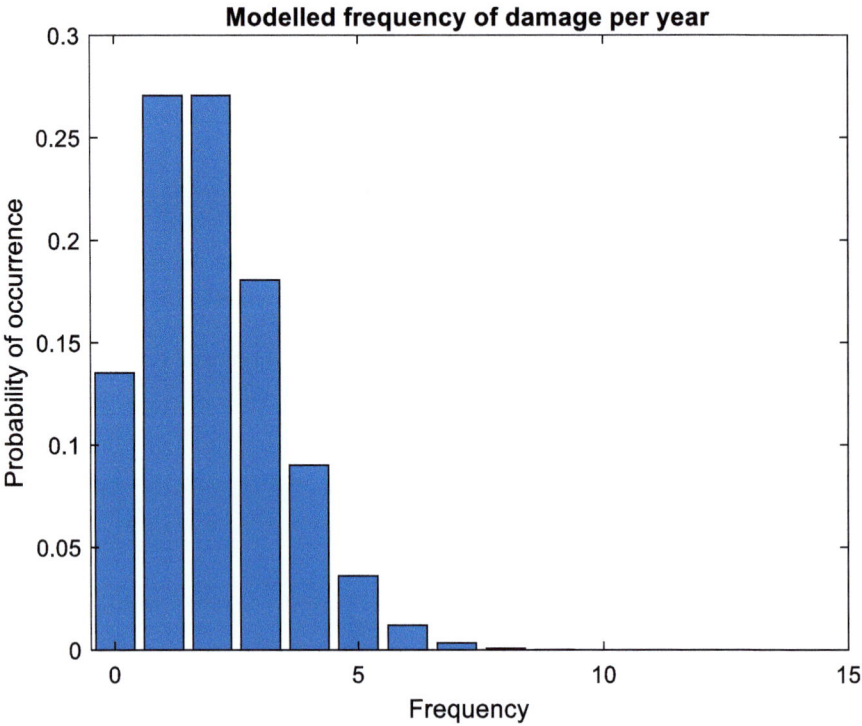

Fig. 4.1 The modelled loss frequency per year

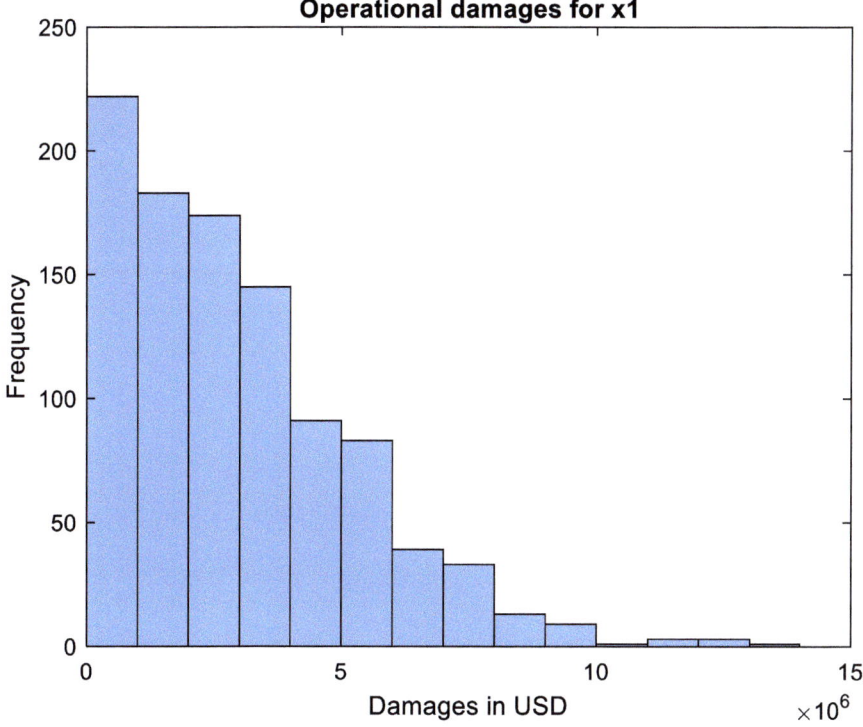

Fig. 4.2 Possible distribution of operational damages

Fig. 4.3 The expected loss and the quantile

		Expected loss	Quantile
1	x1	2917506.64	7248849.70

Literature and Software References

BCBS. (2004). *International Convergence of Capital Measurement and Capital Standards*. ISBN: 92-9197-669-5.
Hull, J. (2018). *Risk Management and Financial Institutions*. 5th ed. Wiley. pp. 515–534.

(continued)

McNeil., A., Frey, R., Embrechts, P. (2015). *Quantitative Risk Management. Concepts, Techniques and Tools*, Princeton University Press, pp. 505–506.

Steinhoff, Carsten (2008). Quantifizierung operationeller Risiken in Kreditinstituten. Cuvillier Verlag

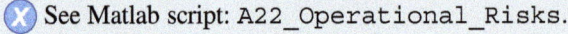 See Matlab script: A22_Operational_Risks.

Chapter 5
Risk Measures

Course Unit 1: Value at Risk-Risk Measures

Assignment 23: Calculating the Value at Risk for a Discrete Probability Distribution

Task
- Calculate the Value at Risk for the MSCI WORLD index price from 12/31t(5) retrospectively for the last 5 years. Proceed with the following steps:
  - First, calculate the Value at Risk for a confidence level of 95% for 1 day based on the discrete returns of the MSCI WORLD index price.
  - Then calculate the Value at Risk for a confidence level of 95% for 1 day based on the MSCI WORLD index price.
- Perform the same calculations for a 99% confidence level.
- Explain the differences in the results and justify your statement.
- Create a chart and show how the Value at Risk can be determined graphically.

Content

> *Value at Risk* (*VaR*) is one of the most important risk measures in financial practice. It describes the amount that, with a given probability, will not be exceeded within a certain period of time. In mathematical terms, the Value at Risk is a quantile of a probability distribution or the distribution of a sample.

(continued)

Supplementary Information The online version contains supplementary material available at https://doi.org/10.1007/978-3-031-42836-4_5.

A one day 99% VaR of 2.5 USD represents a 2.5 USD loss that will not be exceeded with a probability of 99% within 1 day. Subsequently it also means that within 1 day in a maximum of 1% of the cases the loss is greater than 2.5 USD.

In finance, risk is understood as the deviation from the expected value. Therefore, the *Mean Value at Risk* (*MVaR*) is often preferred. The Mean Value at Risk is independent of the expected value. It is obtained by subtracting the expected value from the VaR. In practice, the terms Value at Risk and Mean Value at Risk are often not distinguished, so one should be cautious which term is applicable.

When determining the Value at Risk using historical data, *returns* (relative values) are often more suitable than quantities measured in monetary units (absolute values). The reason for this is that returns tend to have a constant expected value and a constant variance compared to absolute quantities. Suppose the stock market had a price of 107.26 USD at time t(1) and a price of 26.21 USD at time t(2). It can be seen that absolute price fluctuations are higher at a price of 100 USD than at a price of 25 USD. For the percentage changes of the yields, a uniform level can be assumed, respectively.

If the Value at Risk is calculated for a discrete probability distribution, it applies to the chosen *time frame* of the underlying data. If the discrete return is based on daily data, like in this example, then the calculated Value at Risk applies to one day. Scaling up to weekly, monthly, quarterly, or annual values for the Value at Risk is not possible in the presence of a discrete probability distribution. As we will see, this is only possible with a continuous probability distribution.

Important Formulas

Workbook: Case Study Risk Management Worksheet: VaR discr. Return

The Value at Risk corresponds to the mathematical concept of quantiles. The formula for calculating the Value at Risk for a discrete probability distribution is:

$$VaR_{discr,1-\alpha}(X)=Q_{1-\alpha}(X)=\inf\left(x:P(X>x)\leq\alpha\right)=\inf\left(x:P(X\leq x)\geq1-\alpha\right) \quad (5.1)$$

$VaR_{discr,\ 1-\alpha}$	=	Value at Risk at confidence level $1-\alpha$ for discrete distributions
X	=	Random variable which describes the loss function
$Q_{1-\alpha}$	=	$1-\alpha$-quantile of the distribution
$inf(x:\ P(X\leq x)\geq 1-\alpha)$	=	At least $(1-\alpha)\%$ of the values of X are less than or equal to x
$inf(x:\ P(X>x)\leq\alpha)$	=	Maximum $\alpha\%$ of the values of X are greater than x

Note that in a loss function, a loss is positive and a gain is negative. The loss function of the returns is therefore obtained by multiplying the returns by -1.

The Value at Risk of a loss function is the value for which at least $(1-\alpha)\%$ of the values are smaller, i.e. show a lower loss, and a maximum of $\alpha\%$ of the values are larger, i.e. show a higher loss.

Course Unit 1: Value at Risk-Risk Measures

Calculation of the quantile of the distribution to the confidence level $1 - \alpha$ in Excel:

Excel example: `K10=VLOOKUP(K9,F5:H1307,3,0)`

This corresponds to the Value at Risk of the returns given in a discrete probability distribution.

Execution in Excel
- First, the discrete daily returns are calculated from the MSCI WORLD index prices (column D).
- These are then multiplied by -1, since we need a loss function to calculate the Value at Risk (column D).
- The negative returns are then sorted in column H in an ascending order of magnitude.
- To do this, first select the cells G5:H1307 in which the date and the values are noted.
- Then click Data, Sort and Filter, Sort. Enter there as Sort by "column H", Sort on "Cell Values" at Order "Smallest to Largest". Note, we have a loss function, a negative loss is a gain, a positive loss is a loss. Therefore, we sort the returns in such a way that unfavourable results are at the end.
- After that, the confidence level $1 - \alpha$ (cells K6 and N6) and α (cells K7 and N7) are calculated. The confidence level $1 - \alpha$ is the probability with which a loss will not be exceeded and α is the probability with which the values exceed the Value at Risk.
- The next step is to find the value for which at least 95% of the values are smaller, i.e. show a lower loss, and a maximum of 5% of the values are larger, i.e. show a higher loss. With 1303 returns available 95%·1303=1237.85. In this case, the Value at Risk is between the 1237th and 1238th values. Therefore, the corresponding 1238th value of the ordered list in column H must be selected. This is done for the 95% quantile with the function `K9=ROUNDUP((K6)*COUNT(H5:H1307),0)`.
- Subsequently, the return of the 1238th value is determined (sorted in ascending order, starting with the largest return). This is done for the 95% quantile with the function `K10=VLOOKUP(K9,F5:H1307,3,0)`. The determined value is 1.34%. This means that at least 95% of the returns of the given loss function have values less than or equal to 1.34% and at most 5% of the returns of the given distribution function have values greater than 1.34%. To a probability of 95%, within 1 day the loss will not be greater than 1.34%.
- Finally, the Value at Risk is calculated in monetary units `K15=K13*K10`. It amounts to 40.27 USD. Furthermore, the Value at Risk for the number of units is calculated `K19=K15*K17`. This amounts to 201 USD. With a probability of 95%, the loss from the investment will not exceed the value of 201 USD within the next day.

Excel Results (Fig. 5.1)

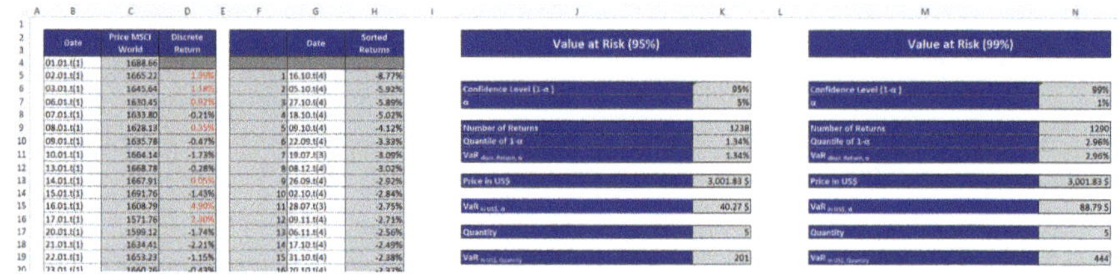

Fig. 5.1 Determination of the VaR for a discrete probability distribution

Execution in Matlab

- Import the required data from the Excel file.

> Data = readmatrix('Matlab Data.xlsx');
> MSCI = Data(:,3);
> Date = Data(:,1);
> Price_MSCI = [Date,MSCI];
> Num_shares = 5; % Number of shares owned

- Calculate the negative discrete returns of the MSCI World. As a loss function is needed to calculate the VaR.

> Return = -(price2ret(MSCI,[],'Periodic'));

- Sort the values in descending order.

> Return = sort(Return);

Course Unit 1: Value at Risk-Risk Measures

- Identify the point where the 95% quantile is exceeded and round it up with ceil().

> Position = ceil(length(Return)*0.95);

- Read the value at this point. The result is 1.34%.

> VaRd1 = Return(Position);

- Alternatively, the point of exceedance can be approximated. This results in a value of 1.35%. The reason for the deviation of the two results lies in the interpolation when calculating the quantile.

> VaRd2 = quantile(Return,0.95);

- Finally, determine the Value at Risk in monetary units and for the number of shares owned.

> VaR_USD = VaRd1*MSCI(end);
> VaR_USD_shares = VaR_USD*Num_shares;

Matlab Results (Fig. 5.2)

	VaRd1	VaRd2	VaR in USD	VaR shares
1	0.0134	0.0135	40.2660	201.3300

Fig. 5.2 Calculated VaR for a discrete distribution in Matlab

Literature and Software References

> Hull, J. (2018). *Risk Management and Financial Institutions*. 5th ed. Wiley. pp. 269–274.
>
> McNeil., A., Frey, R., Embrechts, P. (2015). *Quantitative Risk Management. Concepts, Techniques and Tools*, Princeton University Press, pp. 64–65.
>
> Wewel, Max C.; Blatter, A. (2019): Statistics in undergraduate business and economics. Methods, application, interpretation; with removable collection of formulas. 4th, updated reprint. Munich: Pearson Studium, pp. 185–186.
>
> See Excel file: Case Study Risk Management, Excel worksheet: VaR discr. Return.
> See Matlab script: A23_24_25_VaR_discrete.

Assignment 24: Calculation of the Mean Value at Risk for a Discrete Probability Distribution

Task

- Calculate the Mean Value at Risk for the MSCI WORLD index price from 12/31t(5) retrospectively for the last 5 years. Proceed with the following steps:

  - First, calculate the Mean Value at Risk for a confidence level of 95% for 1 day based on the discrete returns of the MSCI WORLD index price.
  - Then calculate the Mean Value at Risk for a confidence level of 95% for 1 day based on the MSCI WORLD index price

Important Formulas
Workbook: `Case Study Risk Management` Worksheet: `VaR discr. Return`

The formula for calculating the Mean Value at Risk for a discrete probability distribution is:

$$MVaR_{1-\alpha}(X) = VaR_{1-\alpha}(X) - E(X) \quad (5.2)$$

$MVaR_{1-\alpha}$ = Mean Value at Risk at the confidence level $1 - \alpha$ for discrete returns
E = Expected value

Calculate the Mean Value at Risk as a return given a discrete probability distribution in Excel:

Excel example: `K24=ABS(K10-AVERAGE(H5:H1307))`

Execution in Excel
- Based on the previously calculated 95% quantile of 1.34%, the Mean Value at Risk is calculated by subtracting the expected value (mean) of the daily discrete returns of the MSCI WORLD index price from the Value at Risk. The result will be the deviation from the expected value. Since the expected value is 0.05%, the Mean Value at Risk is 1.39% `K24=ABS(K10-AVERAGE(H5:H1307))`.
- For the next step, the Mean Value at Risk in monetary units is calculated by multiplying the Mean Value at Risk for discrete returns by the investment volume. The Mean Value at Risk in monetary units is 41.74 `K26=K24*K13` and for the investment volume 209 USD `K30=K26*K28`.

Excel Results (Fig. 5.3)

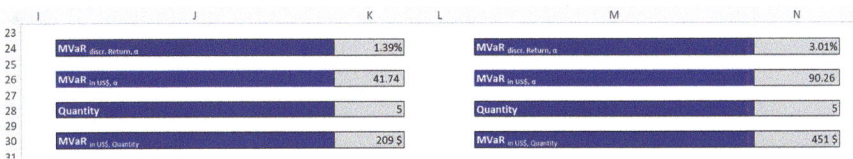

Fig. 5.3 Determination of the Mean VaR for a discrete probability distribution

Execution in Matlab
- Based on the results of the previous assignment, subtract the mean to obtain the Mean VaR (MVaR). Furthermore, calculate the MVaR in monetary units and the number of shares owned.

```
MVaR = VaRd1-mean(Return);
MVaR_USD = MVaR*MSCI(end);
MVaR_USD_shares = MVaR_USD*Num_shares;
```

Matlab Results (Fig. 5.4)

	MVaR	MVaR in USD	MVaR shares
1	0.0139	41.7421	208.7105

Fig. 5.4 Calculated Mean VaR for a discrete distribution in Matlab

Course Unit 1: Value at Risk-Risk Measures

Literature and Software References

> Hull, J. (2018). Risk Management and Financial Institutions. 5th ed. Wiley. pp. 269–274.
> McNeil., A., Frey, R., Embrechts, P. (2015). Quantitative Risk Management. Concepts, Techniques and Tools, Princeton University Press, pp. 64–65.
>
> See Excel file: Case Study Risk Management, Excel worksheet: VaR discr. Return.
> See Matlab script: A23_24_25_VaR_discrete.

Assignment 25: Calculation of the Conditional Value at Risk/Expected Shortfall/Tail Value at Risk for a Discrete Probability Distribution

Task

- Calculate the Conditional Value at Risk for the MSCI WORLD index price from 12/31/t(5) retrospectively for the last 5 years. Proceed by following these steps:
  - First, calculate the Conditional Value at Risk for a confidence level of 95% for 1 day based on the discrete returns of the MSCI WORLD index price.
  - Then calculate the Conditional Value at Risk for a confidence level of 95% for 1 day based on the MSCI WORLD index price.
- Perform the same calculations for a 99% confidence level.
- Explain the differences in the results and justify your statements.
- Create a diagram and show how the Conditional Value at Risk can be determined graphically.

Content

> The *Conditional Value at Risk (CVaR)* or *Expected Shortfall* determines the average amount of losses that exceed the Value at Risk. The Conditional Value at Risk is also referred to as Conditional Tail Expectation or Expected Tail Loss.
>
> The Value at Risk does not provide any information on how high the losses are beyond the limit specified by the Value at Risk. The Conditional Value at Risk overcomes this disadvantage by also including the information (losses) above the Value at Risk. In mathematical terms, the Conditional Value at Risk is a conditional expected value.

(continued)

> Furthermore, in contrast to the Value at Risk, the Conditional Value at Risk satisfies all the conditions for a *coherent risk measure*. In particular, the Conditional Value at Risk is subadditive, a condition not met by VaR.
>
> Value at Risk and Conditional Value at Risk are closely related. While the Value at Risk answers the question of which loss will only be exceeded in at most (100*α)%, the Conditional Value at Risk shows how high the expected loss is in (100*α)% of cases with the highest losses. The CVaR thus indicates, if an unexpected case (of exceeding the value at risk) occurs, how high the expected loss will be.

Important Formulas
Workbook: Case Study Risk Management Worksheet: VaR discr. Return

The formula for calculating the Conditional Value at Risk for returns with a discrete probability distribution is:

$$CVaR_{1-\alpha}(X) = E(X \mid X \geq VaR_{1-\alpha}) \qquad (5.3)$$

$CVaR_{1-\alpha}$ = Conditional Value at Risk at the confidence level $1 - \alpha$
$E(X \mid X \geq VaR_{1-\alpha})$ = Expected loss if the Value at Risk is exceeded

Calculate the Conditional Value at Risk (Expected Shortfall) for returns with a discrete probability distribution in Excel:

> Excel example: K35=AVERAGE(INDIRECT("H"&K9+4&":H1308"))

Calculate the Conditional Value at Risk (Expected Shortfall) in monetary units given a discrete probability distribution in Excel:

> Excel example: K37=K35*K13

Execution in Excel
- Based on the discrete returns sorted upwards in column H, returns greater than or equal to the 95% Value at Risk are selected. The mean value is calculated for these values. The Conditional Value at Risk shows that the expected loss, in the 5% cases with the highest losses, is 2.53% K35=AVERAGE(INDIRECT("H"&K9+4&":H1308")).
- For the next step, the Conditional Value at Risk (Expected Shortfall) is calculated in monetary units. This is 75.87 USD K37=K35*K13. For a number of 5 MSCI WORLD shares, the Conditional Value at Risk in cell K41=K37*K39 is 379 USD.

Course Unit 1: Value at Risk-Risk Measures

Excel Results (Fig. 5.5)

	J	K	L	M	N
34					
35	Conditional Value at Risk discr. Return, α	2.53%		Conditional Value at Risk discr. Return, α	4.88%
36					
37	Conditional Value at Risk in US$, α	75.87 $		Conditional Value at Risk in US$, α	146.39 $
38					
39	Quantity	5		Quantity	5
40					
41	Conditional Value at Risk in US$, Quantity	379 $		Conditional Value at Risk in US$, Quantity	732 $

Fig. 5.5 Determination of the Conditional VaR (Expected Shortfall) for a discrete probability distribution

Execution in Matlab

- The Conditional VaR describes the mean value of the returns that exceed the VaR (1.34%).

```
CVaR = mean(Return(Position:end));
CVaR_USD = CVaR*MSCI(end);
CVaR_USD_shares = CVaR_USD*Num_shares;
```

Matlab Results (Fig. 5.6)

	CVaR	CVaR in USD	CVaR shares
1	0.0253	75.8742	379.3708

Fig. 5.6 Calculated Conditional VaR for a discrete distribution in Matlab

Literature and Software References

Hull, J. (2018). *Risk Management and Financial Institutions.* 5th ed. Wiley. pp. 274.

McNeil., A., Frey, R., Embrechts, P. (2015). *Quantitative Risk Management. Concepts, Techniques and Tools*, Princeton University Press, p. 69.

See Excel file: Case Study Risk Management, Excel worksheet: VaR discr. Return.
See Matlab script: A23_24_25_VaR_discrete.

Assignment 26: Calculation of the Value at Risk with a Continuous Probability Distribution

Task
- Calculate the Value at Risk for the MSCI WORLD index price from 12/31/t(5) retrospectively for the last 5 years. Proceed with the following steps:
  - First, calculate the Value at Risk for a confidence level of 95% for 1 day based on the *continuous* returns of the MSCI WORLD index price.
  - Then calculate the Value at Risk for a confidence level of 95% for 1 day based on the MSCI WORLD index price 3001.83 USD.
  - Calculate the Value at Risk for a confidence level of 99% for 1 day based on the *continuous* returns of the MSCI WORLD index price.
  - Then calculate the Value at Risk for a confidence level of 99% for 1 day based on the MSCI WORLD index price.
- Calculate the previously obtained VaRs also for a period of 30 days.
- Explain the differences in the results and justify your statements.

Content

> For a continuous probability distribution, the calculation of the *Value at Risk* is similar to the calculations for discrete probability distributions. Here, too, the $(1 - \alpha)$-quantile of the distribution is determined, i.e. the value below which at least $(1 - \alpha)$ % *of* the values are. In contrast to the discrete probability distribution, no historical data is used. Instead, a continuous distribution is assumed.
>
> The most common continuous distribution is the *normal distribution*. There, the determination of the quantiles and thus the calculation of the Value at Risk is particularly simple. The normal distribution is defined by the expected value μ and the standard deviation σ. If these parameters are known, the Value at Risk for a given confidence level can be calculated easily. Here we utilise continuous returns.
>
> When calculating the Value at Risk, it is important to note that the *time horizon* of the Value at Risk and the time dimension of the parameters μ and σ must be identical. If the Value at Risk is calculated for one day, as in our example, then also μ and σ as the expected value and standard deviation must be based on daily returns. If the time horizon of the value at risk and the parameters do not match, the expected value μ and the standard deviation σ must be rescaled to the time horizon of the Value at Risk. It is assumed that the parameters μ and σ are constant over time. *Rescaling* is only possible for continuous returns. It is not possible for discrete returns.

Important Formulas

Workbook: Case Study Risk Management Worksheet: VaR continuous return

If the loss function is modelled by a normal distribution with expected value μ and variance σ^2, then the following formula applies for the Value at Risk at the confidence level $(1 - \alpha)$:

$$VaR_{1-\alpha}(X) = \mu + \sigma \cdot N(1-\alpha)^{-1} \qquad (5.4)$$

$VaR_{1-\alpha}(X)$ = Value at Risk at the confidence level $1 - \alpha$ for continuous returns
σ = Standard deviation
$N(1 - \alpha)^{-1}$ = $(1 - \alpha)$-quantile of the standard normal distribution
μ = Expected value

Proof of the correctness of this formula:

$$P(X \leq VaR_{1-\alpha}(X)) = P\left(\frac{X-\mu}{\sigma} \leq N(1-\alpha)^{-1}\right) = N\left(N(1-\alpha)^{-1}\right) = 1-\alpha \quad (5.5)$$

If the Value at Risk is to be calculated for a period other than one day, e.g. for 30 days, the expected value μ and the standard deviation σ must be rescaled to the new time horizon of the Value at Risk. We assume that the parameters refer to one period, in our example one day. Now we want to calculate the Value at Risk for a period with length T, e.g. 30 days. The formula is then:

$$VaR_{1-\alpha}(X) = T \cdot \mu + \sqrt{T} \cdot \sigma \cdot N(1-\alpha)^{-1} \quad (5.6)$$

Calculation of the quantile of the distribution to the confidence level $(1 - \alpha)$ assuming a normal distribution in Excel:

Excel example: H7=NORM.INV(H5,0,1)

Calculation of the continuous daily volatility in Excel:

Excel example: H9=STDEV.P(D5:D1307)

The expected value of the *continuous* returns is calculated in Excel using the function AVERAGE:

Excel example: H11=AVERAGE(D5:D1307)

The Value at Risk is calculated in Excel with:

Excel example: H13=H11+H9*H7

This formula transforms a standard normally distributed random variable into a random variable with a mean μ and standard deviation σ. This can also be done directly using the following Excel command:

Excel example: H13=NORM.INV(H5,H11,H9)

Taking the time dimension into account, the Excel formula for Value at Risk is as follows:

Course Unit 1: Value at Risk-Risk Measures 141

> Excel example: H22=H20*H7*H9+H19*H11

Considering the time dimension, the price and the number of units, the Excel formula for the Value at Risk is as follows:

> Excel example: H27=H23*H25

Execution in Excel
- First, in cell H7=NORM.INV(H5,0,1) the value of the standard normal distribution for the 5% quantile is calculated. It is 1.645.
- Subsequently, the daily volatility is determined H9=STDEV.P(D5:D1307). It amounts to 1%.
- To calculate the Value at Risk, the mean value of the negative, continuous, daily returns is required H11 =AVERAGE(D5:D1307). This has a value of −0.04%.
- The Value at Risk for 1 day is 1.61% and is calculated with the Excel formula: H13=H11+H9*H7. Alternatively via H13 =NORM.INV(H5,H11,H9).
- With a unit price of 3001.83 USD, the Value at Risk in monetary units for 1 day is 48.23 USD. The Excel formula H17=H13*H15 is used.
- Adding the time factor, the Value at Risk is 7.72% in accordance with the Excel formula H22=H20*H7*H9+H19*H11.
- The Value at Risk in monetary units, considering the total investment volume of 5 shares, is 1158 USD. The Excel formula H27=H23*H25 is applied.
- A Value at Risk of 1158 USD means that in 95% of cases the daily loss of an investment in 5 MSCI WORLD Index shares does not exceed 1158 USD.

Excel Results (Fig. 5.7)

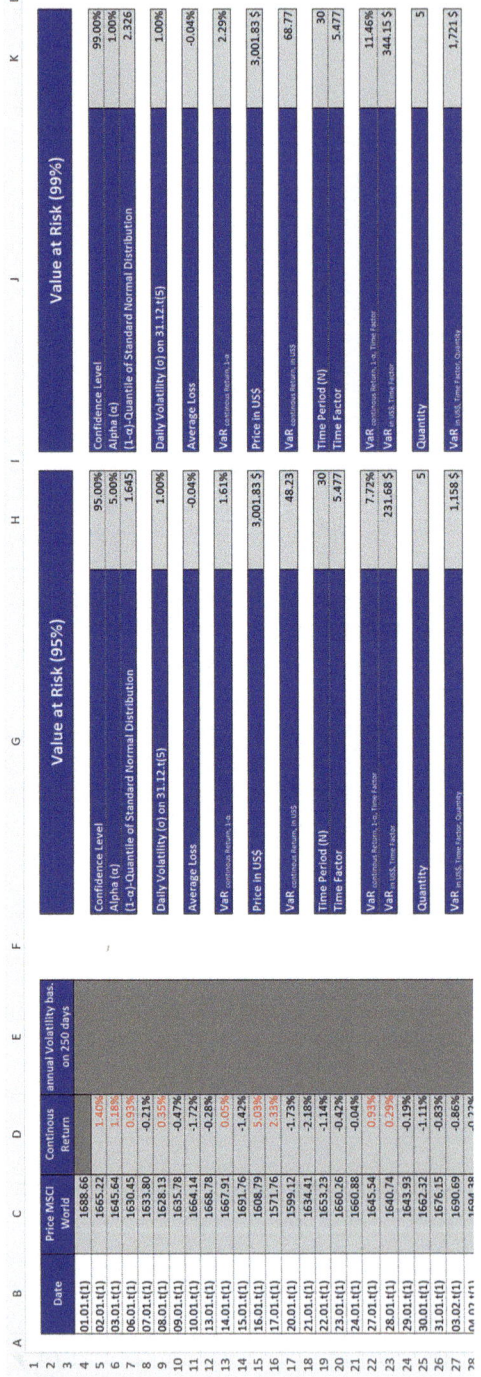

Fig. 5.7 Determining the Value at Risk for a continuous probability distribution

Execution in Matlab
- Import the required data from the Excel file.

```
Data = readmatrix('Matlab Data.xlsx');
MSCI = Data(:,3);
Date = Data(:,1);
Price_MSCI = [Date,MSCI];
```

- Calculate the continuous returns of the MSCI World and denote the losses as positive values.

```
Return = -(price2ret(MSCI,[],'Continuous'));
```

- Identify a suitable normal distribution.

```
pd = fitdist(Return,'Normal');
```

- Read the corresponding value at the 95% quantile (it is 1.61%). Alternatively, the VaR can be calculated manually according to Eq. 5.4.

```
Alpha = 0.05;
VaR = icdf(pd,(1-Alpha));
% Alternative calculation
Expected_value = mean(Return);
Std = std(Return);
Norm_dist = norminv((1-Alpha));
VaR_alternative = Expected_value + Std*Norm_dist;
```

- Determine the VaR in monetary units.

```
MSCI_current_price = MSCI(end);
VaR_USD = VaR*MSCI_current_price;
```

- Determine the VaR with the addition of the time factor.

```
T = 30; % In days
Time_factor = sqrt(T);
VaR_time = T*Expected_value+Time_factor*Std*Norm_dist;
VaR_time_USD = VaR_time*MSCI_current_price;
Num_shares = 5; % Number of shares owned
VaR_time_USD_shares = VaR_time_USD*Num_shares;
```

Matlab Results (Fig. 5.8)

	VaR	VaR in USD	VaR for 30 days	VaR in USD for 30 days	VaR 5 shares in USD for 30 days
1	0.0161	48.2523	0.0772	231.7877	1.1589e+03

Fig. 5.8 Calculated VaRs for a continuous distribution in Matlab

Literature and Software References

> Hull, J. (2018). Risk Management and Financial Institutions. 5th ed. Wiley. pp. 269–274.
> McNeil., A., Frey, R., Embrechts, P. (2015). Quantitative Risk Management. Concepts, Techniques and Tools, Princeton University Press, pp. 64–65.
>
> See Excel file: Case Study Risk Management, Excel worksheet: VaR cont. Return.
> See Matlab script: A26_27_VaR_continuous.

Assignment 27: Calculation of the Conditional Value at Risk or Expected Shortfall for a Continuous Probability Distribution

Task

- Calculate the Conditional Value at Risk for the MSCI WORLD index price from 12/31/t(5) retrospectively for the last 5 years. Proceed with the following steps:
  - First, calculate the Conditional Value at Risk for a confidence level of 95% for 1 day based on the *continuous* returns of the MSCI WORLD index price.
  - Then calculate the Conditional Value at Risk for a confidence level of 95% for 1 day based on the MSCI WORLD index price.
  - Calculate the Conditional Value at Risk for a confidence level of 95% for 1 day MSCI WORLD index.
  - Calculate the Conditional Value at Risk for a confidence level of 99% for 1 day based on the *continuous* returns of the MSCI WORLD index price.
  - Then calculate the Conditional Value at Risk for a confidence level of 99% for 1 day based on the MSCI WORLD index price.
  - Calculate the Conditional Value at Risk for a confidence level of 99% for 30 days based on the MSCI WORLD index price.
- Explain the differences in the results and justify your statements.

Content

> In the case of a continuous probability distribution, the Conditional Value at Risk or Expected Shortfall is the expected value of all losses that are greater than the value at risk. It measures the average loss in the event that the Value at Risk is exceeded.

Important Formulas
Workbook: `Case Study Risk Management` Worksheet: `VaR continuous return`

The Conditional Value at Risk can be calculated using the conditional expected value. The mean is calculated, conditionally that the values exceed the Value at Risk. The conditional expected value corresponds to the calculation of an integral:

$$CVaR_{1-\alpha}(X) = E(X \mid X \geq VaR_{1-\alpha}) \tag{5.7}$$

For a normally distributed random variable $X \sim N(\mu, \sigma^2)$ the Conditional Value at Risk is:

$$CVaR_{1-\alpha}(X) = \sigma \cdot \frac{\varphi\left(N(1-\alpha)^{-1}\right)}{\alpha} + \mu \tag{5.8}$$

$\varphi(\cdot)$ = Density of the standard normal distribution
$N(1-\alpha)^{-1}$ = $1 - \alpha$-quantile of the standard normal distribution
μ = Expected value
σ = Standard deviation

If the Conditional Value at Risk must be calculated for a period other than 1 day, e.g. for 30 days, the expected value μ and the standard deviation σ must be rescaled to the new period.

Substituting the rescaled parameters into the Conditional Value at Risk formula yields the following equation of the rescaled Conditional Value at Risk:

$$CVaR_{1-\alpha}(X) = \sqrt{T} \cdot \sigma \cdot \frac{\varphi\left(N(1-\alpha)^{-1}\right)}{\alpha} + T \cdot \mu \tag{5.9}$$

The value of the density function at the location of the $(1 - \alpha)$ quantile of the standard normal distribution is calculated in Excel as follows:

Excel example: `H32=NORM.DIST(H7,0,1,FALSE)`

Calculation of the Conditional Value at Risk (Expected Shortfall) as return with a continuous probability distribution in Excel:

Excel example: `H33=H11+H9*H32/H6`

Calculation of the Conditional Value at Risk (Expected Shortfall) as return for a period of 30 days with a continuous probability distribution in Excel:

Excel example: H35=$H$20*$H$9*($H$32/$H$6)+$H$19*$H$11

Considering the investment volume, time factor and in monetary units, the Excel formula for the Conditional Value at Risk is as follows:

Excel example: H37=H36*H25

Execution in Excel
- In the first step density function is evaluated at the point of the 95% quantile H32=NORM.DIST(H7,0,1,FALSE). It is 0.1031.
- Then the Conditional Value at Risk for 1 day is calculated H33=H11+H9*H32/H6. The Conditional Value at Risk for 1 day is 2.03%. The CVAR in monetary units is 60.82 USD H34=H15*H33.
- The Conditional Value at Risk for a period of 30 days is calculated in cell H35 H35=$H$20*$H$9*($H$32/$H$6)+$H$19*$H$11. It amounts to 10.02%. The CVAR in monetary units for 30 days is 300.64 USD H36=$H$35*$H$15. Accordingly, for 5 shares it amounts to 1503 USD H37=H36*H25.

Excel Results (Fig. 5.9)

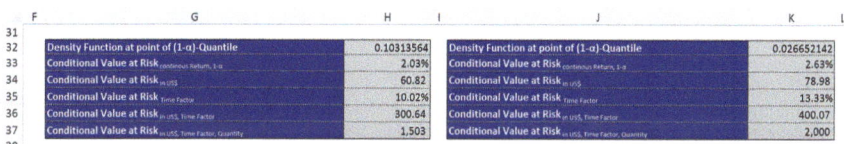

Fig. 5.9 Determination of the Conditional VaR (Expected Shortfall) for a continuous probability distribution

Execution in Matlab

- Starting from Assignment 5.5, Eq. 5.8 must be applied. To do this, multiply the standard deviation by the density of the standard normal distribution at the position of the 95% quantile. Divide the result by alpha and add the expected value to it. This returns a Conditional VaR of 2.03%.

```
Density = normpdf(Norm_dist);
CVaR_return = Std*(Density/Alpha)+Expected_value;
CVaR_time = Time_factor*Std*(Density/Alpha)+T*Expected_value;
CVaR_time_USD = CVaR_time*MSCI_current_price;
CVaR_USD_time_shares = CVaR_time_USD*Num_shares;
```

Matlab Results (Fig. 5.10)

	CVaR	CVaR for 30 days	CVaR in USD for 30 days	CVaR 5 shares in USD for 30 days
1	0.0203	0.1002	300.7717	1.5039e+03

Fig. 5.10 Calculated Conditional VaRs for a continuous distribution in Matlab

Literature and Software References

> Hull, J. (2018). Risk Management and Financial Institutions. 5th ed. Wiley. pp. 269–274.
>
> See Excel file: Case Study Risk Management, Excel worksheet: VaR cont. Return.
> See Matlab script: A26_27_VaR_continuous.

Assignment 28: Backtesting—How Good is the Value at Risk ?

Task

The data set covers 1000 days of data. The confidence level of the VaR is 99%. Therefore, 10 values (1% out of 1000) are expected to exceed the Value at Risk. These values are called "exceptions". The level of significance is expected to be 5%. Assume that:

a) 13 exceptions are observed.
b) 16 exceptions are observed.

You therefore suspect that the Value at Risk underestimates the risk. Check in each case whether you need to reject the Value at Risk model.

Further assume that:

c) 6 exceptions are observed.
d) 4 exceptions are observed.

You suspect that the Value at Risk overestimates the risk. Check in each case whether you need to reject the Value at Risk model.

Content

> The risk models and measures described above are based on assumptions. These assumptions may not accurately represent the reality. It is therefore all of upmost importance to check the (historical) accuracy of the risk measure and models used. One approach to this is backtesting.
>
> *Backtesting* involves checking how well the Value at Risk has predicted the actual, historical risk. If we have a 1-day $VaR_{99\%}$ and have a data set of 1000 days, we expect to observe an exceedance of the VaR on a maximum of 10 days. These exceedances are referred to as "exceptions". It could turn out that in a historical review this value has been exceeded on more days than expected, thus the number of exceptions is higher than predicted.
>
> (continued)

> However, it is questionable at what point the VaR model cannot be used beyond the predicted number of exceptions. Therefore, it must be checked whether the number of exceptions is sufficient to reject the assumptions for the $VaR_{99\%}$ sufficiently. For this purpose, a test of significance known from statistics can be used.

Important Formulas
Workbook: Case Study Risk Management Worksheet: Backtesting

First, the available data is analysed. The data set contains the data from n considered days. In the example above $n = 1.000$. The number of times the Value at Risk is exceeded, the so-called exceptions c are determined from historical data. The probability that such an exception will occur is p and can be calculated by:

$$p = \frac{c}{n} \tag{5.10}$$

$c = $ Number of times the Value at Risk is exceeded, so-called exceptions
$n = $ Number of data in the data set

Consequently, if the $VaR_{1-\alpha}$ is true, the probability that an exception occurs in the data set should be equal to α. Consequently, should hold:

$$p = \alpha \tag{5.11}$$

$p = $ Probability for an exception in the data set
$\alpha = $ Probability for an exception in the VaR model

This statement can be verified with a significance test. In order to perform a significance test, a level of significance must be selected. In practice, 5% is usually used for this purpose. The level of significance means that an error probability of 5% is accepted. It should be noted that in statistical tests a statement can only be rejected and cannot be confirmed.

If the number of observed exceptions c is greater than the number of expected exceptions $\alpha \cdot n$, it must be checked whether the Value at Risk underestimates the risk. The random variable X describes the number of exceptions. One must verify whether:

$$P(X \geq c | p = \alpha) = 1 - P(X < c | p = \alpha) = 1 - P(X \leq c - 1 | p = \alpha) < 5\% \tag{5.12}$$

If the probability that c or more exceptions occur is less than the significance level (here 5%), then the model is rejected. This means that it is unlikely that c or more exceptions can be observed.

If the number of observed exceptions is c is smaller than the number of expected exceptions $\alpha \cdot n$, one must test whether the Value at Risk overestimates the risk. One must verify whether:

$$P(X \leq c \mid p = \alpha) < 5\% \qquad (5.13)$$

If the probability that c or fewer exceptions occur is less than the level of significance, the model is rejected.

The underlying distribution of choice is the binomial distribution, since there are only two possible events: the loss is above the Value at Risk or below. The parameters of the binomial distribution are n, the number of observed days and α, the probability of an exception in the Value at Risk model.

Execution in Excel

In Excel, the value of a binomial distribution can be calculated with BINOM.DIST (Number of Errors, Number of Moves, Alpha, Cumulated).

- First, import the number of trials, alpha, and significance level into cells E4:E6.
- Then enter the number of errors 13 in cell E11.
- In cell E12 we now calculate the value of the distribution with E12=1-BINOM.DIST(E11-1,$E$4,$E$5,TRUE).
- Cell E13 is now used to check whether the VaR is rejected or not. This is done with E13=IF(E12>=E6,"NO","YES").
- In the cells below, repeat this for 16 errors.
- To analyse whether the Value at Risk overestimates the risk, enter the number of errors 6 in cell E22.
- In cell E23 we now calculate the value of the distribution with =BINOM.DIST(E22,$E$4,$E$5,TRUE).
- Cell E24 is now used to check whether the VaR is rejected. This is done E24=IF(E23>=E6,"NO","YES").
- Repeat this for four errors in the cells below.

Excel Results (Figs. 5.11 and 5.12)

	A	B	C	D	E
8					
9		1) Does the Value at Risk underestimate the risk?			
10		a)			
11		Number of exceptions			13
12		Probabilty, to observe x or more exceptions			0.2075
13		VaR is rejected			NO
14		b)			
15		Number of exceptions			16
16		Probabilty, to observe x or more exceptions			0.0479
17		VaR is rejected			YES
18					

Fig. 5.11 Backtesting the VaR in case of underestimation

	A	B	C	D	E
19					
20		**2) Does the Value at Risk overestimate the risk?**			
21		a)			
22		Number of exceptions			6
23		Probabilty, to observe x or less exceptions			0.1289
24		VaR is rejected			NO
25		b)			
26		Number of exceptions			4
27		Probabilty, to observe x or less exceptions			0.0287
28		VaR is rejected			YES
29					

Fig. 5.12 Backtesting the VaR in case of overestimation

Execution in Matlab

In Matlab the function binocdf() can be used to obtain a value of the binominal distribution. In this case the user must specify whether to test for an over- or underestimation.

1) Assumption: VaR underestimates risk

(a) To test for underestimation, use the function 1-binocdf(). The parameters of the binomial distribution are $c - 1 = 12$ exceptions, 1000 trials, and an alpha value of 1-0.99=0.01. The result is $\geq 5\%$. Therefore, the model is not rejected.

```
% With 13 observed exceptions
Y1 = 1- binocdf(12,1000,0.01)
```

(b) To backtest a VaR model with 16 observed exceptions, insert 16-1=15 exceptions into the binocdf() function and proceed as seen above. The results are saved as Y2 and are $\leq 5\%$. Accordingly, the model is rejected.

```
% With 16 observed exceptions
Y2 = 1- binocdf(15,1000,0.01) % <0.05 -> rejected
```

2) Assumption: VaR overestimates risk

(a) To test for overestimation, use the function binocdf(). The parameters of the binomial distribution are 6 exceptions, 1000 trials, and an alpha value of 1-0.99=0.01. The result is $\geq 5\%$. Accordingly, the model is not rejected.

```
% With 6 observed exceptions
X1 = binocdf(6,1000,0.01)
```

(b) To backtest a VaR model with 4 observed exceptions, insert 4 exceptions into the binocdf() function and proceed as seen above. The value is saved as X2. The result is $\leq 5\%$. Hence, the model is rejected.

```
% With 4 observed exceptions
X2 = binocdf(4,1000,0.01) % <0.05 -> rejected
```

Matlab Results (Figs. 5.13 and 5.14)

		13 exceptions	16 exceptions
1	VaR underestimates	0.2075	0.0479

Fig. 5.13 Backtesting the VaR in case of underestimation

		6 exceptions	4 exceptions
1	VaR overestimates	0.1289	0.0287

Fig. 5.14 Backtesting the VaR in case of overestimation

Literature and Software References

> Hull, J. (2018). Risk Management and Financial Institutions. 5th ed. Wiley. pp. 285–288.
> McNeil., A., Frey, R., Embrechts, P. (2015). Quantitative Risk Management. Concepts, Techniques and Tools, Princeton University Press, pp. 352–354.
>
> See Excel file: Case Study Risk Management, Excel worksheet: Backtesting.
> See Matlab script: A28_Backtesting.

Course Unit 2: Lower Partial Moment Risk Measures

Assignment 29: Calculation of Lower Partial Moments— Shortfall Probability

Task

Calculate the shortfall probability (zero-order Lower Partial Moments or LPM_0), for the MSCI WORLD index price from 12/31/t(5) retrospectively for the last 5 years.

Content

> *Lower Partial Moments* (LPM_m -measures) measure downside risk. They capture only the negative deviations from a hurdle τ (target value), but evaluate the entire information of the probability distribution in this range (up to the theoretically possible maximum damage).
>
> Usually, three special cases are considered in practice:
>
> - Shortfall probability (Lower Partial Moments zero order or LPM_0),
> - Shortfall expectation value (Lower Partial Moments first order or LPM_1) and
> - Shortfall variance (Lower Partial Moments second order or LPM_2).
>
> The risk of falling below the target value τ is represented in several ways. The shortfall probability only looks at the probability p of falling below the target. Whereas the expected shortfall value takes the average level of shortfall into account.
>
> The *shortfall probability* (zero-order Lower Partial Moments or LPM_0), also known as downside probability, measures the probability that the minimum requirement level will fall below a predefined threshold. However, it does not capture the absolute level of shortfalls below the target. The shortfall probability or default probability is calculated, for example, as the number of

(continued)

returns that fall short of the target value in relation to the total number of returns. Thus, the shortfall probability represents the relative frequency of undershooting the target value.

In order to calculate the shortfall probability, the minimum target level is set and the probability of falling below this target level is determined. In contrast, in the VaR calculation the confidence level, i.e. the probability of falling below the target, is set and the matching target value is determined. Therefore, the shortfall probability and the VaR are inversely related.

Important Formulas
Workbook: `Case Study Risk Management` Worksheet: `Lower Partial Moments`

The general formula of the Lower Partial Moments is:

$$LPM_n(\tau, r^s) = \frac{1}{T} \sum_{\substack{t=1 \\ X < \tau}}^{T} (\tau - r^s)^n \qquad (5.14)$$

LPM = Lower Partial Moments
τ = Required target value, also "target", e.g. the minimum yield
r^s = Continuous return
t = Index for the time
T = Time span

For the shortfall probability, only the probability P of the shortfall plays a role. The shortfall probability (LPM_0) is calculated based on the above formula as follows:

$$LPM_0(\tau, r^s) = P(r^s < \tau) \qquad (5.15)$$

The calculation of the shortfall probability in Excel is done as follows:

Excel example: `K7=COUNT(E5:E1307)/COUNT(D5:D1307)`

Execution in Excel
- Starting from the continuous returns in column D, the returns X that are below the target value τ (the target value is shown in cell K5 and is -3.5%) are determined. This is obtained by the Excel formula `E5=IF(D5<$K$5,D5,"")`.
- The next step is to calculate the shortfall probability (LPM_0) by dividing the number of values that are below the target by the number of observable returns `K7=COUNT(E5:E1307)/COUNT(D5:D1307)`.

Excel Results (Fig. 5.15)

	A	B	C	D	E	F	G	H	I	J	K
1											
2	Date	Price MSCI World	Continuous Return	=X< τ	=τ-X	=max(τ-X;0)	=max((τ-X)^2;0)		Lower Partial Moment		
3	01.01.t(1)	1688.66									
4	02.01.t(1)	1665.22	-1.40%		-2.10%	0.00%	0.00%		Threshold (Target Value) τ		-3.5%
5	03.01.t(1)	1645.64	-1.18%		-2.32%	0.00%	0.00%				
6	06.01.t(1)	1630.45	-0.93%		-2.57%	0.00%	0.00%		Shortfall-Probability (Lower Partial Moments zeroth Order; LPM₀)		0.7%

Fig. 5.15 Calculation of the shortfall probability (zero-order Lower Partial Moments or LPM$_0$)

The Execution in Matlab can be found in Assignment 5.9.

Literature and Software References

> 📖 Price, K., Price, B., & Nantell, T. J. (1982). Variance and Lower Partial Moment Measures of Systematic Risk: Some Analytical and Empirical Results. The Journal of Finance, 37(3), 843–855.
>
> ❌ See Excel file: Case Study Risk Management, Excel worksheet: Lower Partial Moments.
> See Matlab script: A29_30_31_Lower_Partial_Moments.

Assignment 30: Calculation of Lower Partial Moments— Shortfall Expectation Value

Task

Calculate the shortfall expected value (Lower Partial Moments first order or LPM$_1$) for the MSCI WORLD index price from 12/31/t(5) retrospectively for the last 5 years.

Content

> The *shortfall expected value* (LPM of the first order or LPM$_1$) takes into account the extent to which the target is undershot. It is also referred to as "expected default", "mean default risk", "expected shortfall magnitude" or "target shortfall". At the same time, the expected shortfall can be understood as a special case of the LPM$_1$, in which a moving threshold value equal to the α-quantile is used. Semistandard deviation is also considered as a special case of LPM$_1$.
>
> For the downside expected value, the deviation of the target value and the positive individual returns is calculated and accumulated. This sum is then divided by the total number of returns which are above the threshold. Thus, the

(continued)

Course Unit 2: Lower Partial Moment Risk Measures

LPM$_1$ provides information on the expected level of shortfalls with respect to the target value. Together with the downside probability, this provides a meaningful picture, as it shows not only the probability but also the extent of the possible shortfalls of the target.

Important Formulas
Workbook: `Case Study Risk Management` Worksheet: `Lower Partial Moments`

The shortfall expectation value calculates the average shortfall amount. The calculation of the shortfall expectation value in Excel is done according to:

$$LPM_1(\tau, r^s) = \frac{1}{T} \sum_{\substack{t=1 \\ X < \tau}}^{T} (\tau - r^s) \qquad (5.16)$$

The shortfall expectation value (LPM$_1$) is calculated based on the above formula as follows:

Excel example: `K9=AVERAGE(G5:G1307)`

Execution in Excel
- Starting from the target value τ of -3.5%, the difference between return and target value is measured. This is done using the Excel formula `F5=$K$5-D5`.
- Afterwards those values are determined, which are smaller zero `G5=MAX(F5,0)`.
- In the last step, the shortfall expectation value (LPM$_1$) is determined as the average of the deviations from the bound `K9=AVERAGE(G5:G1307)`.
- The expected value is 0.0179%, indicating that in 0.7% of the cases the losses are on average 0.0179% higher than the specified barrier of -3.5%, or that in 0.7% of the cases the losses are 0.7179%.

Excel Results (Fig. 5.16)

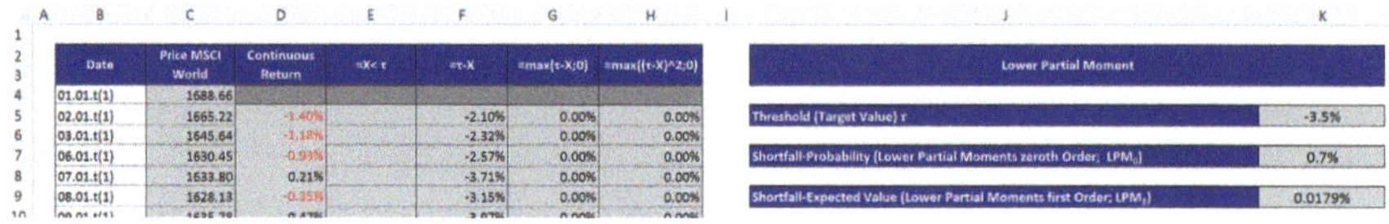

Fig. 5.16 Calculation of the shortfall expectation value (LPM first order or LPM$_1$)

Literature and Software References

Price, K., Price, B., & Nantell, T. J. (1982). Variance and Lower Partial Moment Measures of Systematic Risk: Some Analytical and Empirical Results. The Journal of Finance, 37(3), 843–855.

See Excel file: Case Study Risk Management, Excel worksheet: Lower Partial Moments.
See Matlab script: A29_30_31_Lower_Partial_Moments.

Assignment 31: Calculation of Lower Partial Moments— Shortfall Variance

Task
Calculate the shortfall variance (Lower Partial Moments second order or LPM$_2$) for the MSCI WORLD index price from 12/31 t(5) retrospectively for the last 5 years.

Content

The *shortfall variance* as second-order LPM (LPM$_2$) captures the dispersion of the undershoots of the target value. Using the shortfall variance, larger losses are weighted more heavily than smaller ones by squaring the underruns of the target. The downside variance thus measures the average squared negative deviation from the target value.

The shortfall variance has similarities to the semi-variance, which is why it is sometimes also called target semi-variance. It also has similarities to the variance, although here the deviations are calculated with respect to the target and not with respect to the mean value of the distribution. With the shortfall variance, the high shortfalls of the target value are weighted particularly

(continued)

heavily, whereas the variance penalises strong deviations from the mean in both directions. However, since investors do not want to penalise above-average returns, the shortfall variance fits better with investors' risk preferences.

Important Formulas

Workbook: `Case Study Risk Management` Worksheet: `Lower Partial Moments`

The shortfall variance calculates the dispersion of the undershoots of the target value. The calculation of the shortfall variance in Excel is performed according to:

$$LPM_2(\tau, r^s) = \frac{1}{T} \sum_{\substack{t=1 \\ X<\tau}}^{T} (\tau - r^s)^2 \tag{5.17}$$

The shortfall variance (LPM_2) is calculated based on the above formula as follows:

Excel example: `K11=AVERAGE(H5:H1307)`

Execution in Excel

- Starting from the target value τ of -3.5% cell, the difference between return and target value is measured and squared. This is done by the Excel formula `F5=$K$5-D5`.
- Then those values are determined which are less than zero `G5=MAX(F5,0)` and squared `H5=MAX((G5^2),0)`.
- In the last step, the shortfall variance (LPM_2) is determined as the mean of the squared deviations from the bound `K11=AVERAGE(H5:H1307)`.

Excel Results (Fig. 5.17)

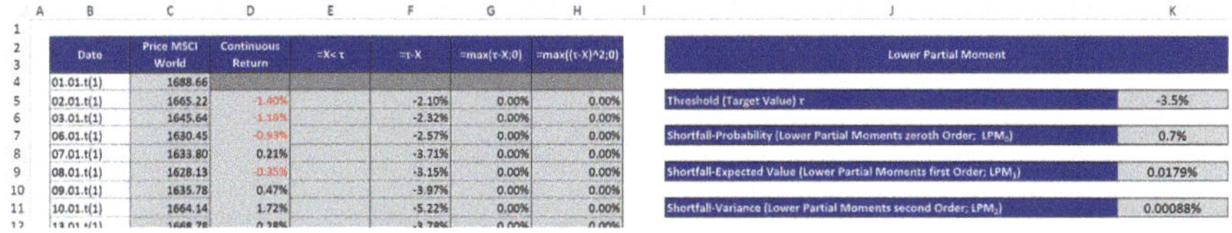

Date	Price MSCI World	Continuous Return	=X< τ	=τ-X	=max(τ-X;0)	=max((τ-X)^2;0)
01.01.t(1)	1688.66					
02.01.t(1)	1665.22	-1.40%		-2.10%	0.00%	0.00%
03.01.t(1)	1645.64	-1.18%		-2.32%	0.00%	0.00%
06.01.t(1)	1630.45	-0.93%		-2.57%	0.00%	0.00%
07.01.t(1)	1633.80	0.21%		-3.71%	0.00%	0.00%
08.01.t(1)	1628.13	-0.35%		-3.15%	0.00%	0.00%
09.01.t(1)	1635.78	0.47%		-3.97%	0.00%	0.00%
10.01.t(1)	1664.14	1.72%		-5.22%	0.00%	0.00%

Lower Partial Moment	
Threshold (Target Value) τ	-3.5%
Shortfall-Probability (Lower Partial Moments zeroth Order; LPM_0)	0.7%
Shortfall-Expected Value (Lower Partial Moments first Order; LPM_1)	0.0179%
Shortfall-Variance (Lower Partial Moments second Order; LPM_2)	0.00088%

Fig. 5.17 Calculation of the shortfall variance as second-order LPM (LPM_2)

Execution in Matlab

- Import the required data from the Excel file.

```
Data = readmatrix('Matlab Data.xlsx');
MSCI = Data(:,3);
Date = Data(:,1);
Price_MSCI = [Date,MSCI];
```

- Calculate the continuous returns of the MSCI World.

```
Continuous_Return = tick2ret(MSCI,[],'continuous');
```

- Determine the threshold of the MSCI World's continuous returns.

```
Threshold = -0.035
```

- Calculate the Lower Partial Moments in % (zero, first and second order).

```
LPM = lpm(Continuous_Return,Threshold,[0 1 2]);
LPM_0_Order = LPM(1)*100;
LPM_1_Order = LPM(2)*100;
LPM_2_Order = LPM(3)*100;
```

Matlab Results (Fig. 5.18)

	LPM 0. order	LPM 1. order	LPM 2. order
1	0.6907	0.0179	8.7655e-04

Fig. 5.18 Calculated Lower Partial Moments in percent

heavily, whereas the variance penalises strong deviations from the mean in both directions. However, since investors do not want to penalise above-average returns, the shortfall variance fits better with investors' risk preferences.

Important Formulas
Workbook: Case Study Risk Management Worksheet: Lower Partial Moments

The shortfall variance calculates the dispersion of the undershoots of the target value. The calculation of the shortfall variance in Excel is performed according to:

$$LPM_2(\tau, r^s) = \frac{1}{T} \sum_{\substack{t=1 \\ X < \tau}}^{T} (\tau - r^s)^2 \qquad (5.17)$$

The shortfall variance (LPM_2) is calculated based on the above formula as follows:

Excel example: K11=AVERAGE(H5:H1307)

Execution in Excel
- Starting from the target value τ of -3.5% cell, the difference between return and target value is measured and squared. This is done by the Excel formula F5=$K$5-D5.
- Then those values are determined which are less than zero G5=MAX(F5,0) and squared H5=MAX((G5^2),0).
- In the last step, the shortfall variance (LPM_2) is determined as the mean of the squared deviations from the bound K11=AVERAGE(H5:H1307).

Excel Results (Fig. 5.17)

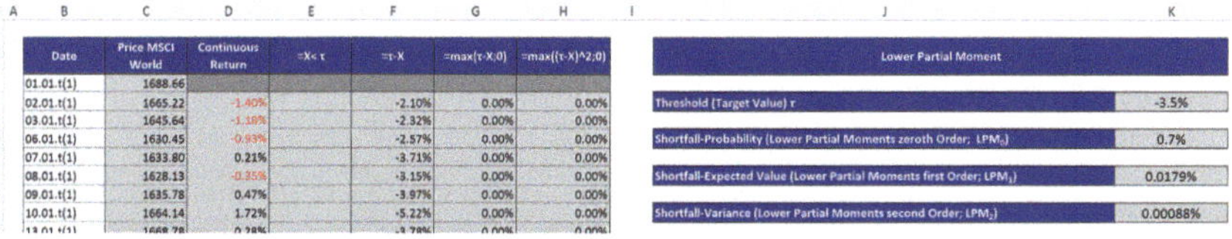

Fig. 5.17 Calculation of the shortfall variance as second-order LPM (LPM_2)

Execution in Matlab
- Import the required data from the Excel file.

```
Data = readmatrix('Matlab Data.xlsx');
MSCI = Data(:,3);
Date = Data(:,1);
Price_MSCI = [Date,MSCI];
```

- Calculate the continuous returns of the MSCI World.

```
Continuous_Return = tick2ret(MSCI,[],'continuous');
```

- Determine the threshold of the MSCI World's continuous returns.

```
Threshold = -0.035
```

- Calculate the Lower Partial Moments in % (zero, first and second order).

```
LPM = lpm(Continuous_Return,Threshold,[0 1 2]);
LPM_0_Order = LPM(1)*100;
LPM_1_Order = LPM(2)*100;
LPM_2_Order = LPM(3)*100;
```

Matlab Results (Fig. 5.18)

	LPM 0. order	LPM 1. order	LPM 2. order
1	0.6907	0.0179	8.7655e-04

Fig. 5.18 Calculated Lower Partial Moments in percent

Literature and Software References

> Price, K., Price, B., & Nantell, T. J. (1982). Variance and Lower Partial Moment Measures of Systematic Risk: Some Analytical and Empirical Results. The Journal of Finance, 37(3), 843–855.
>
> See Excel file: `Case Study Risk Management`, Excel worksheet: `Lower Partial Moments`.
> See Matlab script: `A29_30_31_Lower_Partial_Moments`.

Course Unit 3: Bond Risk Measures, Extreme Risks and Risk Measures in Comparison

Assignment 32: Macaulay Duration and Modified Duration

Task

You have a bond with a remaining term of 6 years. The nominal interest rate is 4.5%. The face value of the bond (amount which is repaid at the end of the term) is 1000 USD.

- Calculate the present value of the payments at a market interest rate of 4%.
- Calculate the Macaulay Duration.
- Calculate the Modified Duration.
- Estimate the price of the bond due to the change in the market interest rate to 5% using the concept of Modified Duration. Compare this approximated new price to the actual new price.
- Calculate the Convexity.

Content

> *Bonds* are debt securities which are emitted to the capital market. These financial instruments are vital for professional investors such as banks and insurances due to their capability of attracting large sums of capital (a benchmark financial bond emission usually is sized at around half a billion USD). We assume that you have mastered the basics of bonds (see references).
> The following basic concepts are relevant:

Bond	Usually longer-term debt security. The issuer pays the purchaser of the bond interest (coupons) for the agreed term, and at the end of the term nominal amount is paid in full (in practice also referred to as bullet repayment)
Coupon	The coupon is the nominal payment the investor receives from the bond. In the case of a fixed rate bond the coupon can be broken down into two components: 1. The average market interest rate until maturity (yield-to-maturity), the value of which is derived from forward payments 2. The credit spread of the issuer (indicating the Probability of Default of the issuer) A plain-vanilla floating rate bond theoretically only has the credit spread component as it matches its interest level with the next coupon payment. However, there is a certain inter-correlation between spreads and interest rates as a higher interest rate environment can increase the likelihood of a default
Nominal value/ Face value	Amount of money that the issuer of the bond must repay to the holder of the bond at the end of the term
Nominal interest rate	Contractually agreed interest rate at which interest is paid on the bond
Fixed rate bond	Bond with interest rate fixed in advance. Interest is usually paid annually or semi-annually

The *Duration*, or *Macaulay Duration*, indicates the point in time at which the value of the fixed-income security is independent of how the market interest rate has changed. This is possible because changes in interest rates have opposite effects on the final value of a fixed-income security. On the one hand, a rise in interest rates results in a lower present value of the security, while on the other hand, coupon payments can earn a higher interest rate when reinvested.

The concept of duration was introduced by Frederick R. Macaulay. The Macaulay Duration can also be interpreted as the average capital commitment period of an investment in a fixed-income security. It is the weighted average of the times at which the investor receives payments from a security. The Macaulay Duration thus indicates the time after which the investor will recover his invested capital and thus describes the reinvestment risk. The concept is mostly applied to fix rate bonds as the duration of a floating rate bond corresponds to the time of the next coupon payment at which the interest of the bond will adapt to current market levels. In practice these bonds are often bought by conservative investors such as banks, insurances and pension funds. To eliminate the interest rate risk, these investors may choose to transform fixed rate bonds into floating rate bonds with asset Swaps.

Modified Duration is a risk measure for interest rate risk and can be derived from the duration (hence Modified Duration). While the duration is measured in years, Modified Duration measures the relative change in the bond price

(continued)

depending on a change in the market interest rate level. Modified Duration indicates the percentage by which the bond price changes if the market interest rate level changes by one percentage point. It thus indicates how much the total return on a bond (consisting of the redemptions, coupon payments and the compound interest effect when the redemptions are reinvested) changes if the interest rate on the market changes. Or in other words, the Modified Duration measures the price effect triggered by a marginal interest rate change and can be expressed as the elasticity of the bond price as a function of the market interest rate.

The present value of bonds shows a convex course in the case of interest rate changes. Since the Modified Duration only considers the first derivative—i.e. the slope—it only provides usable values for small changes in interest rates. Modified Duration is a very cautious risk measure, as risks are overestimated and opportunities underestimated. Convexity is therefore used as a somewhat more accurate risk measure. With Convexity, not only the first derivative is considered, but also the second derivative.

The *Convexity* is a risk measure used to describe the behaviour of a bond when interest rates change. It is an extension or improvement of modified duration. Positive convexity describes bonds that have low price sensitivity when interest rates are rising and high price sensitivity when interest rates are falling. When interest rates are rising, low price losses can be expected, but when interest rates are falling, high price increases can be expected. Furthermore, the greater the convexity, the more pronounced this bond behaviour is.

Important Formulas

Workbook: Case Study Risk Management Worksheet: Duration

The *Macaulay Duration* can be calculated using the following formula. For simplification reasons, a flat yield curve with an annually constant market interest rate is assumed.

$$D = \frac{\sum_{t=1}^{T} t \cdot \frac{Z_t}{(1+r)^t}}{P} = \frac{\sum_{t=1}^{T} t \cdot \frac{Z_t}{(1+r)^t}}{\sum_{t=1}^{T} \frac{Z_t}{(1+r)^t}} \qquad (5.18)$$

D = Macaulay Duration
Z_t = Payment at the end of period t
r = Market interest rate
P = Price of the bond
t = Time period
T = Term of the bond

If the effect of the change in the market interest rate on the price of a bond shall be analysed, the price function of the bond is differentiated with respect to the market interest rate. The following relationship is obtained:

$$\frac{dP(r)}{dr} = -\frac{1}{1+r} \cdot D \cdot P(r) \tag{5.19}$$

$\frac{dP(r)}{dr}$ = First derivative of the price function according to the market interest rate
$dP(r)$ = Change in price in the event of a small change in the market interest rate
dr = Small change in the market interest rate

Transforming this equation yields:

$$\frac{dP}{P} = -\frac{1}{1+r} \cdot D \cdot dr = -MD \cdot dr \tag{5.20}$$

As a simplification, it is $P(r) = P$. The modified Macaulay Duration is the proportionality factor:

$$MD = -\frac{1}{1+r} \cdot D \tag{5.21}$$

Modified Duration is therefore nothing more than the first derivative of the present value function according to the interest rate divided by the price (present value) of the bond.

The concept of Convexity improves the results of the Duration. This is achieved by adding the second derivative in the Taylor expansion. The following relationship can be derived:

$$P(r + \Delta r) \approx P(r) + \frac{dP}{dr} \cdot \Delta r + \frac{1}{2!} \cdot \frac{d^2P}{dr^2} (\Delta r)^2 \tag{5.22}$$

Δr = Change in the market interest rate
$\frac{d^2P}{dr^2}$ = Second derivative of the price function according to the market interest rate

and thus:

$$\frac{P(r + \Delta r) - P(r)}{P(r)} = \frac{\Delta P}{P} \approx -MD \cdot \Delta r + \frac{1}{2} \cdot C \cdot (\Delta r)^2 \tag{5.23}$$

where C describes the Convexity:

$$C = \frac{\Delta^2 P}{d\Delta r^2} \cdot \frac{1}{P} = \frac{\sum_{t=1}^{T} t \cdot (t+1) \frac{Z_t}{(1+r)^{t+2}}}{P} = \frac{\sum_{t=1}^{T} t \cdot (t+1) \frac{Z_t}{(1+r)^{t+2}}}{\sum_{t=1}^{T} \frac{Z_t}{(1+r)^t}} \quad (5.24)$$

Convexity is the second derivative of the present value function with respect to the interest rate divided by the price (present value) of the bond.

The formula for the price change after the exact calculation is:

$$\Delta P = P(r + \Delta r) - P(r) \quad (5.25)$$

After an interest rate change, the new bond price must be recalculated using a present value calculation. As an approximation, however, the price change can be calculated using the concept of Modified Duration.

The formula for the price change after the approximate calculation via the Modified Duration is:

$$P(r + \Delta r) - P(r) = 1/(1+r) \cdot D \cdot P(r) \cdot \Delta r = MD \cdot P(r) \cdot \Delta r \quad (5.26)$$

The formula for the price change after the approximate calculation via Convexity is:

$$P(r + \Delta r) - P(r) = MD \cdot \Delta r \cdot P(r) + \frac{1}{2} \cdot C \cdot (\Delta r)^2 \cdot P(r) \quad (5.27)$$

Execution in Excel

The payment series (C16 to H16) of the bond consists of coupon payments (calculated via the nominal value of the bond (C4) and the nominal interest rate (C8)). At the end of the term, the nominal value is repaid (C4). Please note, with the prefix "$" the formula can be copied to the right without changing the cells.

```
Excel example:
C16=$C$4*$C$8
D16=$C$4*$C$8
E16=$C$4*$C$8
F16=$C$4*$C$8
G16=$C$4*$C$8
H16=$C$4*$C$8+C4
```

With this payment series, the present value can now be calculated easily by discounting the individual values and then adding them up:

Excel example:
C19=C$16*(1+$C$9)^(-C$15)
...
H19=H$16*(1+$C$9)^(-H$15)
I19=SUM(C19:H19)

The sum corresponds to the present value of the payments. The present value of the payments is also the price of the bond.

The numerator of the Duration formula is a weighted present value. First, the individual present values are weighted by time t and then added up:

Excel example:
C21=C$15*C19
...
H21=H$15*H19
I21=SUM(C21:H21)

Macaulay Duration is now obtained by dividing weighted present value and present value:

Excel example: C28=I21/I19

To calculate the Modified Duration, the Macaulay Duration must still be divided by the interest factor:

Excel example: C32=-C28/(1+C9)

For the calculation of the Macaulay Duration and the Modified Duration Excel offers predefined functions:

The calculation of the Duration is done with the Excel function DURATION according to:

Excel example: C51=DURATION(C43,C44,C45,C46,C48,C49)

The calculation of the Modified Duration is done with the Excel function MDURATION according to:

Excel example: C52=-MDURATION(C43,C44,C45,C46,C48,C49)

The calculation of the weighted present value in the numerator relevant for Convexity is performed in Excel according to:

```
Excel example:
C23=C15*(1+C15)*C16/((1+$C$9)^(C15+2))
...
H23=H15*(1+H15)*H16/((1+$C$9)^(H15+2))
I23=SUM(C23:H23)
```

The weighted present value for Convexity is then linked to cell C36. In order to calculate the Convexity, this should be divided by the present value, which we have linked to cell C37.

The calculation of Convexity is done in Excel according to:

```
Excel example: C39=C36/C37
```

The calculation of the total present value of all payments after increasing the market interest rate is performed in Excel in the same way as the above present value calculation, except that the present value must be calculated using the new market interest rate:

```
Excel example:
C55=C$16*(1+$C$11)^(-C$15)
...
H55=H$16*(1+$C$11)^(-H$15)
I55=SUM(C55:H55)
```

The exact calculation of the price change is done in Excel according to:

```
Excel example: C62=C58-I19
```

The approximate calculation of the price change using the Modified Duration is done in Excel according to:

Excel example: C66=C32*C10*I19

The approximate calculation of the price change using Convexity is done in Excel according to:

Excel example: C67=C66+1/2*C39*C10^2*I19

Execution in Excel
- In the payments (cells C16:H16) resulting from the bond, you can see the coupon/nominal interest rate. In our example, 1000 USD has been invested. The interest payments are 45 USD per year, i.e. the coupon/nominal interest rate is 4.5% (cell C8).
- Based on the payments, the present values of the payments are calculated using the current market interest rate (C9) and the market interest rate after interest rate increase (C11) (cells C19 to I19 and C55 to I55). The present value of the payments discounted with the current interest rate (I19) corresponds to the price/rate of the bond.
- In the next step, the Duration, Modified Duration and Convexity of the bond are calculated without and with Excel functions.
- To calculate the Macaulay Duration, the weighted present value of the Duration (cells C21:H21) is calculated in the numerator. The already calculated present values (C19:H19) are weighted with the corresponding time periods (C15:H15). In cell I21, the weighted present value of Duration is determined as the sum of the individual weighted values.

    This present value is divided by the price of the bond (I19), from which the Macaulay Duration is obtained (cell C28). The same result is obtained by the Excel function DURATION (cell C51), which requires the inputs from cells C43:C49 (except C47).

    Both calculations result in a Macaulay Duration of 5.399, which means that the investor will recover his invested money after 5.399 years.
- The Modified Duration of the bond is calculated in cell C32. It results from the negative value of the Macaulay Duration divided by the market yield plus 1 (cell C9 +1). The same result is obtained by the Excel function MDURATION (cell C52).

    Both calculations result in a Modified Duration of -5.1914, which means that the price of the bond decreases by 5.1914 USD if the market interest rate increases by 1% (assuming that the bond price would depend linearly on the interest rate). However, this assumption is only an approximation, since the interest rate and bond price behave convexly.
- Therefore, we use Convexity in the next step. To calculate it, the weighted present value of Convexity (cells C23:H23) is calculated in the numerator.

This present value (I23) is divided by the price of the bond (I19), resulting in the Convexity (cell C39). An Excel function for calculating the Convexity is not available.

The calculation of Convexity yields a value of 33.6791. The Convexity stands for the curvature of the curve of the bond price. It shows how a change in interest rates affects the Duration. Since we have a positive Convexity here, we can conclude for this bond that it has low price sensitivity when interest rates are rising and high price sensitivity when interest rates are falling. This is important additional information when we compare bonds with the same or similar Modified Duration.

- The next step is to determine the price change, i.e. the loss if the interest rate increases. This can be done via an exact calculation or an approximate calculation.
- In the exact calculation, the difference between the present value before and after the interest rate increase is calculated (cell C62).
- The approximate calculation of the price change using Modified Duration is C66=C32*C10*I19. It results in a price change of −53.27 USD. This shows that using the Modified Duration overestimates the risk.
- We get a better result when using Convexity
  C67=C66+1/2*C39*C10^2*I19. Here the price change is −51.55 USD, which is very close to the exact value of −51.59 USD.
- You will certainly ask yourself why we make approximate calculations for the price change when we can get the desired result right away with the simpler, exact calculation. Especially if we have several securities with non-linear functions in the portfolio, we rely on the approximate calculation.

Excel Results (Figs. 5.19, 5.20 and 5.21)

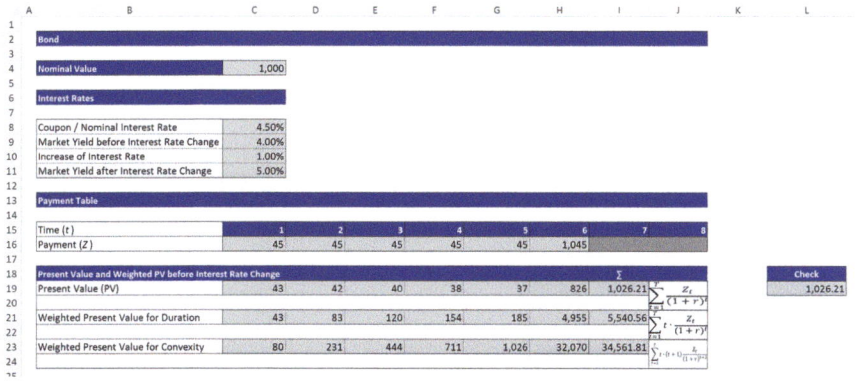

Fig. 5.19 Calculation of interest rates and present values

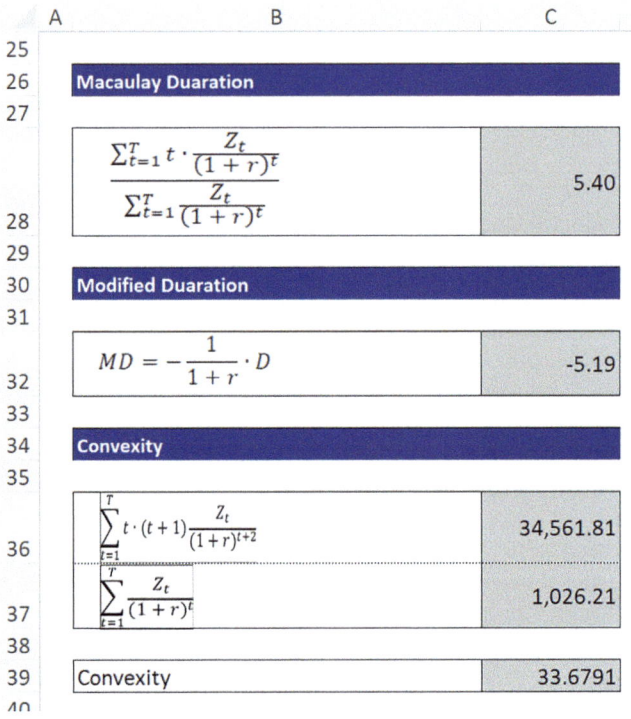

Fig. 5.20 Calculation of Duration, Modified Duration and Convexity

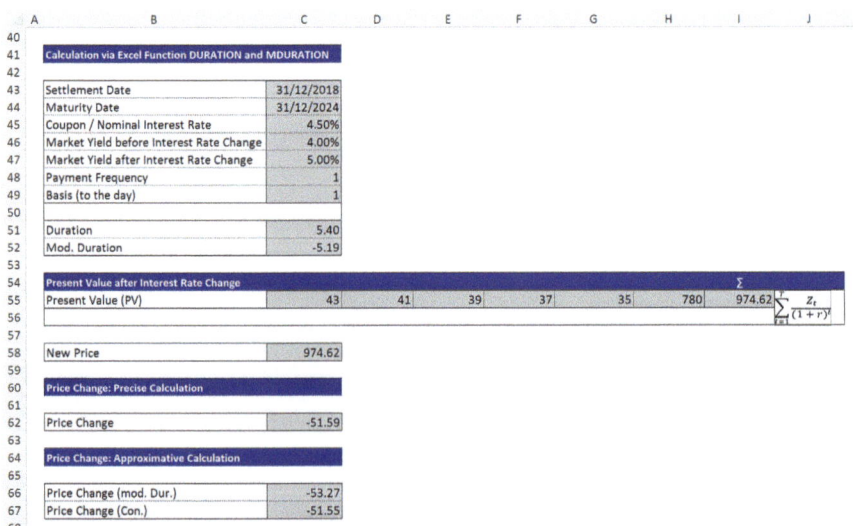

Fig. 5.21 Calculation of Duration and Modified Duration with Excel functions and calculation of price changes

Execution in Matlab
- Define the following assumptions of the bond.

```
Nominal_value = 1000;
Nominal_interest_rate = 0.045;
Market_rate = 0.04;
Cupon = Nominal_value*Nominal_interest_rate;
Rate_hike = 0.01;
Interest_rate_new = Market_rate + Rate_hike;
```

- Determine the payment series of the bond.

```
Payments = [0 Cupon Cupon Cupon Cupon Cupon (Cupon +Nominal_value)].';
```

- Calculate the present values of the bond for the current market interest rate and after the rate hike.

```
PV_market_rate = pvvar([Payments],Market_rate)
PV_rate_hike = pvvar([Payments],Interest_rate_new)
```

- Calculate the Macaulay and Modified Duration.

```
[MacaulayDuration, ModifiedDuration] = cfdur(Payments(2: end), Market_rate)
```

- Calculate the Convexity of the bond.

```
Convexity = cfconv(Payments(2: end),Market_rate)
```

Matlab Results (Fig. 5.22)

	PV market rate	PV rate hike	Macaulay Duration	Modified Duration	Convexity
1	1026.21	974.62	5.40	-5.19	33.68

Fig. 5.22 Calculation of the present values before and after the interest rate hike, the Duration, the Modified Duration and the Convexity in Matlab

Literature and Software References

> 📘 Hull, John (2018): Options, futures, and other derivatives. 9th ed. Essex: Global Edition. pp. 114–117.
>
> ❌ See Excel file: `Case Study Risk Management`, Excel worksheet: `Duration`.
> See Matlab script: `A32_Duration`.

Assignment 33: Extreme Value Theory

Task
- Using Extreme Value Theory, calculate the Value at Risk and Conditional Value at Risk (Expected Shortfall) for 1 day at a confidence level of 95% for the MSCI WORLD index price from 12/31 backward for the last 5 years.
- Perform a sensitivity analysis and determine the Value at Risk and Conditional Value at Risk for the following confidence levels: 99.0%; 99.1%; 99.2%; 99.3%; 99.4%; 99.5%; 99.6%; 99.7%; 99.8%; 99.9%.

Content

> The concept of extreme events (e.g., disruptive technologies, financial market crises, etc.), as described, for example, by *Taleb* in his *"Black Swans"*, considers events that cannot be predicted (or that are difficult to foresee) and that may have serious economic consequences. This phenomenon is usually not taken into account in traditional risk management. In practical risk management and for dealing with the uncertainty of the future, it seems reasonable to use known historical data to infer *extreme values* in the future. Here, techniques are needed that avoid the use of a normal distribution. Instead it should be noted that extreme events occur much more frequently than implied by a normal distribution. The techniques required here are based on the fractal

(continued)

geometry of Mandelbrot and rely essentially on the so-called scalable distributions, such as the *Pareto distribution*. This is the origin of the Extreme Value Theory, which we apply in the following.

The *Extreme Value Theory (EVT)*, here following the Peaks-over-Threshold approach (PoT), goes beyond the expected shortfall, which is calculated on the basis of available historical data. Extreme Value Theory is used for potentially catastrophic events that occur very seldomly but produce extremely high loss amounts. Extreme Value Theory provides a scientific approach to predicting these rare events. It is a mathematical discipline that deals with outliers.

Similar to the expected shortfall Extreme Value Theory deals with the tails of distributions. It takes into account the empirical observations that extreme returns have a higher probability in reality (so-called *fat tails*) and mean returns have a lower probability than described by a normal distribution. Extreme Value Theory can be used to accurately calculate values for high confidence levels. Extreme events occur rarely and can be described as events exceeding a threshold value.

Extreme Value Theory can explain both the left and right tails of a distribution. Here, we deal with the left tail of the distribution.

When using the Extreme Value Theory, it should be noted that the Pareto distribution is only valid for "extreme" events, i.e. for events above a threshold determined by the user. The threshold should be set at a sufficiently high level such that only the extreme tail of the distribution is examined. Conversely, a too large threshold limits the number of exceedances and leads to a high variance.

Important Formulas
Workbook: Case Study Risk Management Worksheet: Extreme Value Theory

The (cumulative) generalised Pareto distribution is calculated with the following equation:

$$G_{\xi,\beta}(y) = 1 - \left(1 + \xi \frac{y}{\beta}\right)^{-1/\xi} \tag{5.28}$$

ξ = Shape parameter of the Pareto distribution
β = Scaling parameter of the Pareto distribution

The distribution has two parameters, ξ (with $\xi \neq 0$)) and β. These must be estimated. The parameter ξ is a shape parameter that determines the severity of the edge of the distribution. The parameter β is a scaling parameter.

The parameters ξ and β can be estimated using the Maximum-Likelihood-Method. For this purpose, the returns are sorted by size. Let $\hat{r}$ be the critical threshold. For the choice of the critical threshold, a value close to the 95% quantile is recommended. Only returns above the critical threshold are considered, i.e. $r_i > \hat{r}$.

The likelihood function to be maximised is:

$$\prod_{i=1}^{n_{\hat{r}}} \frac{1}{\beta}\left(1 + \frac{\xi(r_i - \hat{r})}{\beta}\right)^{-\frac{1}{\xi}-1} \tag{5.29}$$

$\hat{r}$ = Critical threshold
r_i = Values above the critical threshold
$n_{\hat{r}}$ = Number of values above the critical threshold

The maximisation of this function corresponds to the maximisation of its logarithm. The following representation follows from the calculation rules of the logarithm:

$$\sum_{i}^{n_{\hat{r}}} \ln\left[\frac{1}{\beta}\left(1 + \frac{\xi(r_i - \hat{r})}{\beta}\right)^{-\frac{1}{\xi}-1}\right] \tag{5.30}$$

Calculation of the log-likelihood function in Excel:

Excel example: J12=IF(H12>$G$5, LN((1/$N$4)*((1+($N$5*($H12-$G$5)/$N$4)))^(-1/$N$5-1)),0)

Using the iterative search method (SOLVER in Excel), we can now search for the parameters ξ and β that maximise the previous formula.

After determining the parameters ξ and β the Value at Risk can be calculated. The formula is:

$$VaR = \hat{r} + \frac{\beta}{\xi}\left[\left(\frac{n}{n_{\hat{r}}}(1-p)\right)^{-\xi} - 1\right] \tag{5.31}$$

n is the number of observations, $n_{\hat{r}}$ is the number of values below the threshold, and p is the confidence level.

Calculation of the Value at Risk in Excel:

Excel example: J5=G5+(N4/N5)*(((J2/J3)*(1-J4))^(-N5)-1)

The conditional Value at Risk is calculated using the following formula:

$$CVaR = \frac{VaR + \beta - \xi \hat{r}}{1 - \xi} \quad (5.32)$$

Calculation of the Conditional Value at Risk in Excel:

Excel example: J6=(J5+$N$4-$G$5*$N$5)/(1-$N$5)

Execution in Excel
- First, the continuous returns, as a loss function, are calculated in column D.
- The continuous daily returns are then sorted by size in a descending order in column H using the Sort function.
- After that, the confidence level p is set to determine the threshold value $\hat{r}$ is set (cell G2) and the alpha α is calculated (cell G3) G3=1-G2 is set.
- In the next step, the number of α-% smallest values is determined. In our example, it is the 5% smallest values. With 1303 available returns, this is 65 values. Here the function G4=ROUNDDOWN((G3)*COUNT(D12:D1314),0) is applied.
- Subsequently, the yield of the 65th smallest value is determined. This is done for the 5% quantile with the function G5=MAX(VLOOKUP(G4,G12:H1314,2,0),0). The calculated rate of return is 1.379%. The threshold value $\hat{r}$ is 1.379%.
- Column J contains the calculation of the probability for the Maximum-Likelihood-Method. The Excel formula is: J12=IF(H12>$G$5,LN((1/$N$4)*((1+($N$5*($H12-$G$5)/$N$4)))^(-1/$N$5-1)),0).
- To calculate the likelihood for the Maximum-Likelihood-Method, we need, in addition to the threshold value $\hat{r}$ the values of the parameters ξ and β. The initial values for determining ξ and β are first obtained from the Assumptions and are linked to M4 and M5. We link this cell in turn with the cells N4 and N5, in which, after optimisation with the solver, the values of β and ξ will stand.
- In cell J8, the probabilities are summed up J8=SUM(J12:J1314), which are then maximised during optimisation.
- For optimisation using the SOLVER, the following values must be entered into the SOLVER (Fig. 5.23).
- This results in values for β in the amount of 0.008442 and for ξ in the amount of 0.3212.
- In the next step, the Value at Risk can be calculated using the Extreme Value Theory. For this purpose, the following variables are required in addition to β and ξ also the number of observations n J2=COUNT(J12:J1314) the number of values n_r that lie above the threshold J3=COUNTIF(J12:J1314,"<>0") as well as the confidence level p and the threshold value $\hat{r}$ (cell G5). The VaR is calculated as =G5+(N4/N5)*(((J2/J3)*(1-J4))^(-N5)-1). It is 1.36% and, in contrast to the previous calculations of the Value at Risk, describes the exact value at the 5% quantile.

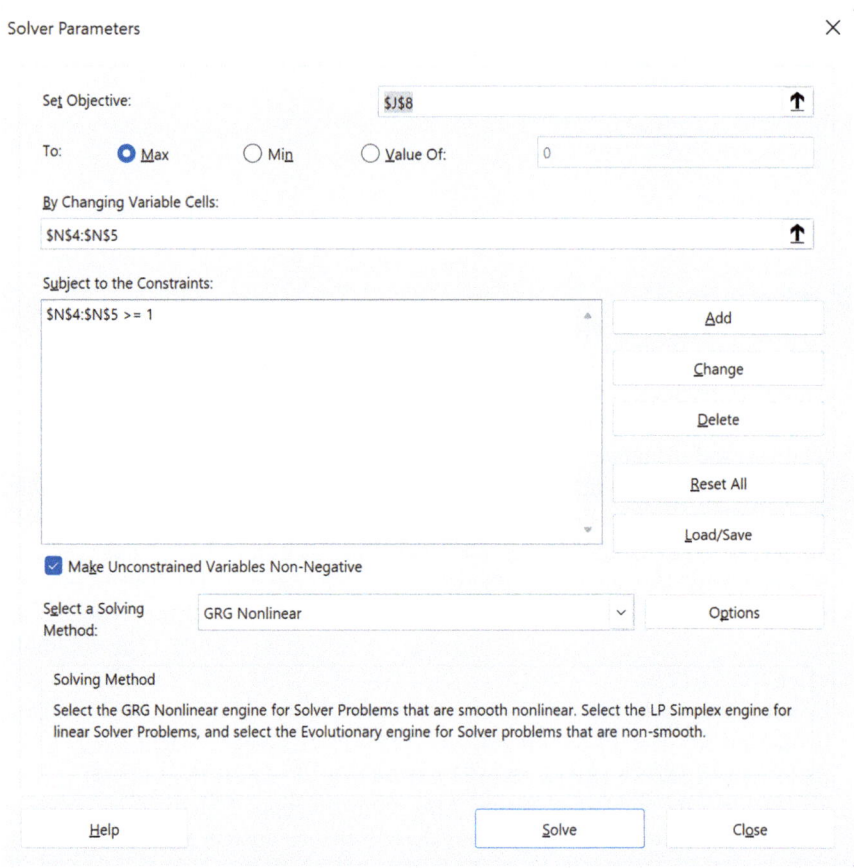

Fig. 5.23 Solver parameters for the Extreme Value Theory

- Furthermore, the Conditional Value at Risk can be calculated with the Extreme Value Theory `J6=(J5+$N$4-$G$5*$N$5)/(1-$N$5)`. It amounts to 2.60%.

Excel Results (Figs. 5.24 and 5.25)

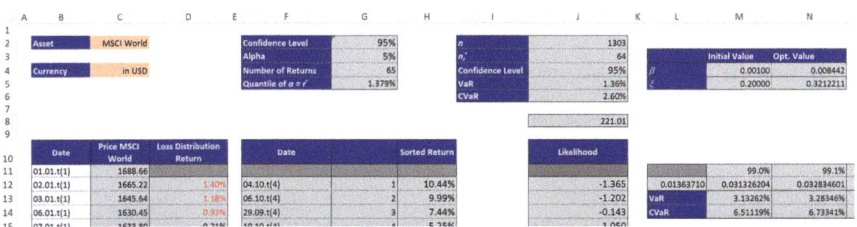

Fig. 5.24 Calculation of the VaR and Conditional Value at Risk with the Extreme Value Theory

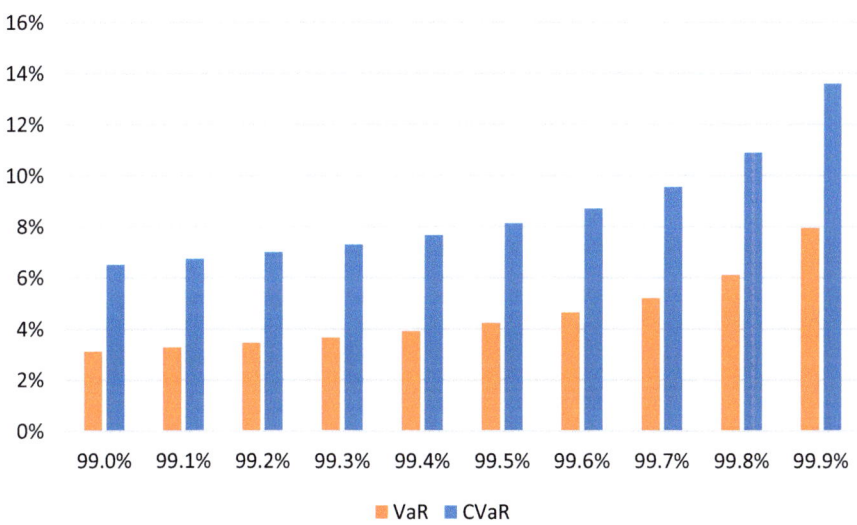

Fig. 5.25 VaR and Conditional Value at Risk at different confidence levels

Execution in Matlab
- Import the required data from the Excel file.

```
Data = readmatrix('Matlab Data.xlsx');
MSCI = Data(:,3);
Date = Data(:,1);
Price_MSCI = [Date,MSCI];
```

- Calculate and sort the negative continuous returns of the MSCI World.

```
Continuous_Return = -price2ret(MSCI,[],'Continuous');
Sorted_Return = sort(Continuous_Return,'descend');
```

- Define the following assumed parameters.

```
Confidence_Level = 0.95;
Alpha = 1-Confidence_Level;
hat = [0.2 0.001];
Xi_hat = hat(1)
Beta_hat = hat(2)
```

- Determine the threshold value $\alpha = \widehat{r}$.

```
Num_Return_Alpha = floor(Alpha* length(Continuous_Return));
Quantil_Alpha = Sorted_Return(Num_Return_Alpha) % Threshold
```

- Calculate the probabilities for the Maximum-Likelihood-Method.

```
Return_Threshold = Sorted_Return(Sorted_Return > Quantil_Alpha);
Likelihood_Hat    =    log((1/Beta_hat)*((1+(Xi_hat*(Return_Threshold-
Quantil_Alpha)/Beta_hat))).^(-1/Xi_hat-1));
LogLikelihood_Hat = sum(Likelihood_Hat)
```

- Optimise the assumed parameters by minimisation.

```
fun = @(hat)sum(-log((1/hat(2))*((1+(hat(1)*(Return_Threshold-Quantil_Alpha)/
hat(2)))).^(-1/hat(1)-1)));
[Opt,SummeLogLikelihood,exitflag] = fminsearch(fun,[Xi_hat Beta_hat]); %
Max-Likeli-Method
Likelihood_Opt = -(SummeLogLikelihood)
Xi_opt = Opt(1)
Beta_opt = Opt(2)
```

- Calculate the Value at Risk and the Expected Shortfall.

```
Num_Return_Total = length(Sorted_Return);
Num_Return_Threshold = length(Return_Threshold(Return_Threshold >
Quantil_Alpha));
Value_at_Risk = (Quantil_Alpha+(Beta_opt/Xi_opt)*(((Num_Return_Total/
Num_Return_Threshold)*Alpha).^(-Xi_opt)-1))
Expected_Shortfall = (Value_at_Risk+Beta_opt-Quantil_Alpha*Xi_opt)/
(1-Xi_opt)
table(Value_at_Risk,Expected_Shortfall,'VariableNames',{'Value    at
Risk','Expected Shortfall'})
```

Matlab Results (Fig. 5.26)

	Value at Risk	Expected Shortfall
1	0.0136	0.0260

Fig. 5.26 Calculation of the VaR and Conditional Value at Risk with the Extreme Value Theory

Literature and Software References

Embrechts, P., Klüppelberg, C., & Mikosch, T. (1997): Modelling extremal events for insurance and finance. Springer, 352–370.

Embrechts, P., Resnick, S. I., & Samorodnitsky, G. (1999): Extreme Value Theory as a Risk Management Tool. North American Actuarial Journal, 3(2), 30–41.

See Excel file: Case Study Risk Management, Excel worksheet: Extreme Value Theory.

See Matlab script: A33_ Extreme_Value_Theory.

Assignment 34: Risk Measures in Comparison

Task
- Review the risk measure requirements listed in the following table.
- Fill out the table.
- Discuss your decisions (Fig. 5.27).

Request	Variance	Standard deviation	Semi-variance/semi-standard deviation	Absolute Value at Risk	Mean Value at Risk	Conditional Value at Risk	LPM$_0$	LPM$_1$	LPM$_2$
Easy interpretability									
Possibility of direct measurement of economic risk									
Use as a target for optimization problems									
Possibility of integrated risk measurement of different risk types									
Use for risk management of a portfolio									
Coherence									

Fig. 5.27 Risk measures in comparison

Content

In general, *risk indicators* and *risk measures* are used to quantify risks and, on this basis, to implement control measures. The aim of quantifying risks using risk measures is to identify risks that could jeopardise the company's existence and to counteract them.

To quantify the risk, a risk measure must be used that adequately reflects the level of risk. In general, risk can be represented in the form of a distribution function of a random variable, as we have already discussed. However, this form of representing a risk is often not very meaningful and comprehensible for non-experts, so that a condensation of the information is desirable.

As we have already seen with the risk measures we use, the level of risk is described by the extent of deviation from the expected value. For risk management purposes, risk measures should allow statements to be made about the probability of occurrence, on the one hand, and about the amount of loss, on the other. In addition, the selected risk measure should be easy to understand and interpret. It is therefore advisable to express risk in monetary units. At first sight, the Value at Risk meets these requirements.

In summary, a risk measure should meet the following requirements:

- Ease of interpretation,
- Possibility of direct measurement of economic risk,
- Use as a target for optimisation problems,
- Possibility of integrated risk measurement of different risk types,
- Use for risk management of a bank portfolio as well as
- Coherence.

Coherent Risk Measures (Table 5.1)

Table 5.1 Coherent risk measures according to Artzner, Delbaen, Eber and Heath (1999)

	Mathematical condition	Financial interpretation
Monotony	If $Z_1 \leq Z_2$ then $p(Z_1) \leq p(Z_2)$	If the risk Z_1 is lower than the risk Z_2, then the capital requirement should also be lower
Subadditivity	$p(Z_1 + Z_2) \leq p(Z_1) + p(Z_2)$	The risk of the sum of the sub-portfolios is less than or equal to the sum of their individual risks. In terms of a portfolio, this means: combining risks diversifies the portfolio
Positive homogeneity	If $\alpha \geq 0$, then $p(\alpha Z) = \alpha\, p(Z)$	When you double a position, you double the risk
Translation invariance	If α is a real number, then $p(Z + \alpha) = p(Z) - \alpha$	Adding a safe amount, reduces the risk

Chapter 6
Aggregation

Assignment 35: Variance–Covariance Method: Variance–Covariance Matrix and Portfolio Risk

Task

Create a portfolio consisting of the MSCI World Index and bonds with coupon payments equal to the three months EURIBOR +1%. The assets of the portfolio consist of 80% MSCI World and 20% bonds.

- Calculate the daily continuous returns for the assets in the portfolio.
- Calculate the average returns for each portfolio asset.
- Calculate the covariances for the assets in the portfolio and create a variance–covariance matrix.
- Calculate portfolio returns based on historical portfolio asset returns and portfolio weights.
- Calculate the portfolio variance and portfolio standard deviation based on the variance–covariance matrix and portfolio weights.

Content

> The *variance–covariance method* belongs to the analytical and parametric methods. The term "parametric method" is derived from the fact that parameters such as the standard deviation are used for the calculation. Under the simplifying assumptions of the parametric approach, the expected value and the standard deviation are sufficient to determine the Value at Risk.

(continued)

Supplementary Information The online version contains supplementary material available at https://doi.org/10.1007/978-3-031-42836-4_6.

> In the variance–covariance method, the returns of the various assets in the portfolio are assumed to have a joint normal distribution. Since the expected values of the individual returns are known due to the available empirical data, the portfolio return can be calculated easily. Also, the standard deviations of the individual returns as well as the covariances can be determined, so that the variance–covariance matrix can be used to calculate the portfolio risk.
>
> Using the portfolio return and the portfolio standard deviation, we can now derive the Value at Risk, the Mean Value at Risk and the Conditional Value at Risk for normally distributed data.

Important Formulas

Workbook: Case Study Risk Management Worksheet: Variance-Covariance Method

The expected return on the portfolio is calculated as:

$$\mu_P = \sum_{i=1}^{m} w_i \cdot \mu_i \tag{6.1}$$

μ_i = Expected value of r_i
w_i = Weight of the asset i
m = Number of assets

The formula for calculating the mean of the historical return, which is the expected value of the return, is:

> Excel example: J14=AVERAGE(D15:D1317)

The formula of the covariance of a sample is:

$$Cov[r_{i,j}] = \sigma_{i,j} = \frac{1}{n-1} \sum_{t=1}^{n} (r_{i,t} - \mu_i) \cdot (r_{j,t} - \mu_j) \tag{6.2}$$

μ_i = Expected return of asset i, expected value of r_i
$r_{i,t}$ = Return at time t of asset i

The covariance in the variance–covariance matrix is calculated as follows:

> Excel example: J17=COVARIANCE.S(D15:D1317,D15:D1317)

Assignment 35: Variance–Covariance Method: Variance–Covariance...

Portfolio risk measured as portfolio variance is calculated using the following formula:

$$\sigma_P^2 = \sum_{i,j=1}^{m} w_i \cdot w_j \cdot \sigma_i \cdot \sigma_j \cdot \rho_{i,j} = \sum_{i,j=1}^{m} w_i \cdot w_j \cdot \sigma_{i,j} \quad (6.3)$$

w_i = Weight of the asset i
σ_i = Volatility of the asset i
$\rho_{i,j}$ = Correlation coefficient between asset i and asset j
$\sigma_{i,j}$ = Covariance matrix between asset i and asset j

The portfolio variance is obtained as follows:

Excel example: J27=MMULT(MMULT(TRANSPOSE(J21:J22),J17:K18),J21:J22)

The portfolio standard deviation is the root of the portfolio variance:

Excel example: J28=SQRT(J27)

Execution in Excel

The initial situation is from the MSCI WORLD index price (column C) and the bond (column F). The MSCI World represents the market risk, the bond the credit risk.

- The next step is to calculate the mean values of the historical returns, which serve as expected values J14=AVERAGE(D15:D1317).
- The basis for the calculation of the portfolio variance is the variance–covariance matrix, which is calculated as follows: J17=COVARIANCE.S(D15:D1317, D15:D1317).
- Furthermore, the weights of the individual assets in the portfolio are 0.8 MSCI WORLD 0.2 bond. Finally, the portfolio return, portfolio variance and portfolio standard deviation are calculated. The portfolio return is calculated as J26=@MMULT(J14:K14,J21:J22) the portfolio variance is calculated as J27=MMULT(MMULT(TRANSPOSE(J21:J22),J17:K18),J21:J22) and the portfolio standard deviation is calculated as J28=SQRT(J27).

Excel Results (Fig. 6.1)

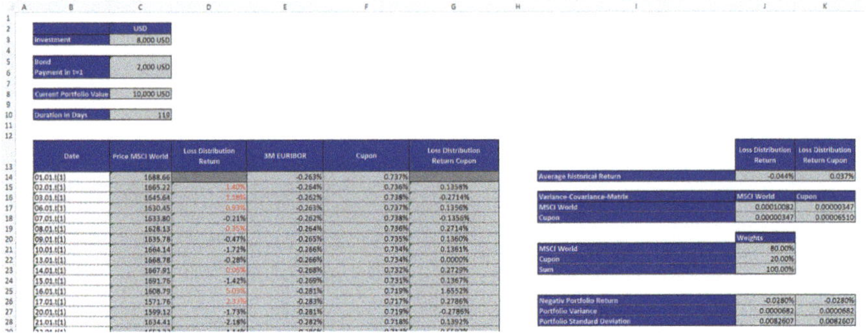

	USD
Investment	8,000 USD
Bond Payment in t=1	2,000 USD
Current Portfolio Value	10,000 USD
Duration in Days	119

Date	Price MSCI World	Loss Distribution Return	3M EURIBOR	Cupon	Loss Distribution Return Cupon
01.01.t(1)	1688.66		-0.263%	0.737%	
02.01.t(1)	1665.22	1.40%	-0.264%	0.736%	0.1358%
03.01.t(1)	1645.64	1.18%	-0.262%	0.738%	-0.2714%
06.01.t(1)	1630.45	0.93%	-0.263%	0.737%	0.1356%
07.01.t(1)	1633.80	-0.21%	-0.262%	0.738%	-0.1356%
08.01.t(1)	1628.13	0.35%	-0.264%	0.736%	0.2714%
09.01.t(1)	1635.78	-0.47%	-0.265%	0.735%	0.1360%
10.01.t(1)	1664.14	-1.72%	-0.266%	0.734%	0.1361%
13.01.t(1)	1668.78	-0.28%	-0.266%	0.734%	0.0000%
14.01.t(1)	1667.91	0.05%	-0.268%	0.732%	0.2729%
15.01.t(1)	1691.76	-1.42%	-0.269%	0.731%	0.1367%
16.01.t(1)	1608.79	5.03%	-0.281%	0.719%	1.6552%
17.01.t(1)	1571.76	2.37%	-0.283%	0.717%	0.2786%
20.01.t(1)	1599.12	-1.73%	-0.281%	0.719%	-0.2786%
21.01.t(1)	1634.41	-2.18%	-0.282%	0.718%	0.1392%

	Loss Distribution Return	Loss Distribution Return Cupon
Average historical Return	-0.044%	0.037%

Variance-Covariance-Matrix	MSCI World	Cupon
MSCI World	0.00010082	0.00000347
Cupon	0.00000347	0.00006510

	Weights
MSCI World	80.00%
Cupon	20.00%
Sum	100.00%

Negativ Portfolio Return	-0.0280%	-0.0280%
Portfolio Variance	0.0000682	0.0000682
Portfolio Standard Deviation	0.0082607	0.0082607

Fig. 6.1 Creation of the variance–covariance matrix and calculation of the portfolio risk using the variance–covariance method

Execution in Matlab

- Import the required data of the MSCI World as well as the 3M EURIBOR.

```
Data = readmatrix('Matlab Data.xlsx');
MSCI = Data(:,3);
ThreeM_EURIBOR = Data(:,2);
Coupon = ThreeM_EURIBOR+1;
Date = Data(:,1);
EURIBOR = [Date,ThreeM_EURIBOR];
Price_MSCI = [Date,MSCI];
```

- Calculate the continuous returns of the coupon payments and the MSCI World.

```
Return_Coupon = log(Coupon(1:end-1)./(Coupon(2:end)));
Return_MSCI = -price2ret(MSCI);
```

- Create a weighted portfolio of both assets.

```
Weight = [.8.2]'; % 80% MSCI 20% Coupon
Portfolio = [Return_MSCI Return_Coupon];
Portfolio_Return = (Portfolio*Weight);
Mean_Return = mean(Portfolio_Return);
```

- Calculate the variance, covariance and standard deviation.

```
Covariance = cov(Portfolio);
Portfolio_Variance = (Weight'*Covariance)*Weight;
Portfolio_Std = sqrt(Portfolio_Variance);
```

Matlab Results (Figs. 6.2 and 6.3)

Fig. 6.2 The calculated covariance matrix

	Covariance	
1	1.0082e-04	3.4718e-06
2	3.4718e-06	6.5097e-05

	Portfolio Variance	Portfolio Standard Deviation
1	6.8239e-05	0.0083

Fig. 6.3 The calculated portfolio variance and standard deviation

Literature and Software References

Hull, J. (2018). Risk Management and Financial Institutions. 5th ed. Wiley. p. 264.

McNeil., A., Frey, R., Embrechts, P. (2015). Quantitative Risk Management. Concepts, Techniques and Tools, Princeton University Press, pp. 340–341.

See Excel file: Case Study Risk Management, Excel worksheet: Variance-Covariance-Method.

See Matlab script: A35_36_Var_Covariance.

Assignment 36: Variance–Covariance Method: Calculation of the Value at Risk and Conditional Value at Risk

Task
- Using the calculated portfolio return and standard deviation, create a density function assuming a normal distribution.
- Calculate the portfolio Value at Risk for a confidence level of 95% and for a maturity of one 1 day/30 days based on the portfolio volume in USD.
- Calculate the Portfolio-Mean-Value at Risk (Deviation Portfolio Value at Risk) for a maturity of 30 days based on the portfolio volume in USD.
- Calculate the Conditional Portfolio Value at Risk (Expected Portfolio Shortfall) for a maturity of 1 day/30 days based on the portfolio volume in USD.
- Plot the Portfolio Value at Risk, Mean Portfolio Value at Risk, and Conditional Portfolio Value at Risk on the density function graph.

Content

Since the variance–covariance method assumes a joint normal distribution for the continuous returns of the various assets in the portfolio, the Value at Risk and the Conditional Value at Risk can be calculated for normally distributed data using the portfolio return and the portfolio standard deviation. The calculation method does not differ from the methods used for the Value at Risk and Conditional Value at Risk for a normal distribution.

Important Formulas

Workbook: Case Study Risk Management Worksheet: Variance–Covariance Method

The formula determining the Value at Risk is:

Assignment 36: Variance–Covariance Method: Calculation of the Value...

$$VaR_{1-\alpha}(X) = \sqrt{T} \cdot \sigma_t \cdot N(1-\alpha)^{-1} + T \cdot \mu_t \tag{6.4}$$

$VaR_{1-\alpha}(X)$	=	Value at Risk at the confidence level $1 - \alpha$
σ_t	=	Standard deviation
$N(1-\alpha)^{-1}$	=	$1-\alpha$-quantile of the standard normal distribution
μ_t	=	Expected value of the portfolio loss
T	=	Maturity

Calculation of the quantile of the distribution to the confidence level $1 - \alpha$ assuming a normal distribution in Excel:

Excel example: J33=NORM.INV(J30,0,1)

The Value at Risk is calculated in Excel with:

Excel example: J36=J26+J28*J33

For a given portfolio volume, the formula of the Value at Risk in monetary units for one day is:

Excel example: J40=J36*J38

Adding the time factor results in a Value at Risk of 1.33% according to the Excel formula:

Excel example: J45=J28*J33*J43+J26*J42

The conditional Value at Risk is calculated as:

$$CVaR_{1-\alpha} = \sqrt{T} \cdot \sigma_t \cdot \frac{\varphi\left(N(1-\alpha)^{-1}\right)}{\alpha} + T \cdot \mu_t \tag{6.5}$$

$\varphi(\cdot)$	=	Density of the standard normal distribution
σ_t	=	Standard deviation
$N(1-\alpha)^{(-1)}$	=	$1-\alpha$-quantile of the standard normal distribution
μ_t	=	Expected value of the portfolio loss
T	=	Maturity

$\varphi(\cdot)$ is the density of the standard normal distribution.

For the Conditional Value at Risk, the density is calculated in Excel as follows:

Excel example: J51=NORMDIST(J33,0,1,FALSE)

Calculation of the Conditional Value at Risk considering the time factor with a continuous probability distribution in Excel:

Excel example: J52=J43*J28*J51/J31+J42*J26

Calculation of the Conditional Value at Risk considering the time factor in monetary units with a continuous probability distribution in Excel:

Excel example: J53=J38*J52

Execution in Excel
- When calculating the Value at Risk, the value of the standard normal distribution for the 5% quantile is first calculated in cell J33=NORM.INV(J30,0,1). It is 1.645.
- The Value at Risk for one day is 1.33% and is calculated with the Excel formula J36=J26+J28*J33.
- For a portfolio volume of 10,000 USD, the Value at Risk in monetary units for one day is 133 USD. The Excel formula J40=J36*J38 is used.
- For a Time Period of 30 days, the Value at Risk is 6.6% according to the Excel formula K45=K28*K33*K43+K26*K42.
- The corresponding Value at Risk in monetary units is 660 USD. The Excel formula K46=K38*K45 will be used.
- Value at Risk means that in 95% of the cases, within the next 30 days, the loss from an investment in the portfolio will not exceed 660 USD.
- When calculating the Conditional Value at Risk, the first step is to calculate the density for the 95% quantile K51=NORMDIST(K33,0,1,FALSE)). It is 0.1031.
- The Conditional Value at Risk for 30 days is then calculated as follows: K52=K43*K28*K51/K31+K42*K26. The Conditional Value at Risk for 30 days is 8.49%.
- The Conditional Value at Risk is converted into monetary units by multiplying the conditional Value at Risk for 30 days by the portfolio volume K53=K38*K52. The Conditional Value at Risk in monetary units is 849 USD.
- If the Value at Risk is exceeded, the Conditional Value at Risk states that the expected loss is 849 USD.

Excel Results (Fig. 6.4)

Execution in Matlab

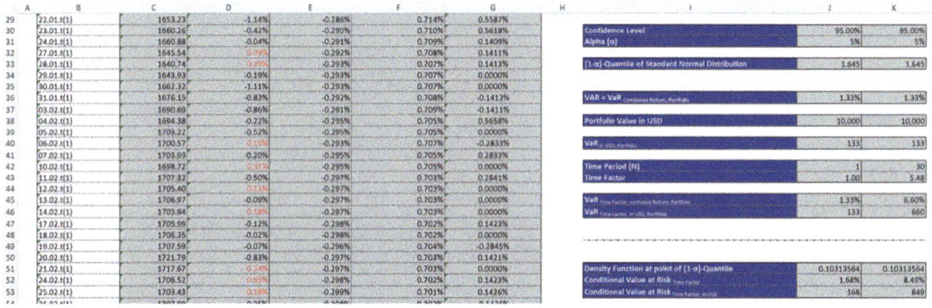

	A	B	C	D	E	F	G
29		22.01.t(1)	1653.23	-1.14%	-0.286%	0.714%	0.5587%
30		23.01.t(1)	1660.26	-0.42%	-0.290%	0.710%	0.5618%
31		24.01.t(1)	1660.88	-0.04%	-0.291%	0.709%	0.1409%
32		27.01.t(1)	1645.54	0.93%	-0.292%	0.708%	0.1411%
33		28.01.t(1)	1640.74	0.29%	-0.293%	0.707%	0.1413%
34		29.01.t(1)	1643.93	-0.19%	-0.293%	0.707%	0.0000%
35		30.01.t(1)	1662.32	-1.11%	-0.293%	0.707%	0.0000%
36		31.01.t(1)	1676.15	-0.83%	-0.292%	0.708%	-0.1413%
37		03.02.t(1)	1690.69	-0.86%	-0.291%	0.709%	-0.1411%
38		04.02.t(1)	1694.38	-0.22%	-0.295%	0.705%	0.5658%
39		05.02.t(1)	1703.22	-0.52%	-0.295%	0.705%	0.0000%
40		06.02.t(1)	1700.57	0.16%	-0.293%	0.707%	-0.2833%
41		07.02.t(1)	1703.93	-0.20%	-0.295%	0.705%	0.2833%
42		10.02.t(1)	1698.72	0.31%	-0.295%	0.705%	0.0000%
43		11.02.t(1)	1707.32	-0.50%	-0.297%	0.703%	0.2841%
44		12.02.t(1)	1705.40	0.11%	-0.297%	0.703%	0.0000%
45		13.02.t(1)	1706.97	-0.09%	-0.297%	0.703%	0.0000%
46		14.02.t(1)	1703.94	0.18%	-0.297%	0.703%	0.0000%
47		17.02.t(1)	1705.99	-0.12%	-0.298%	0.702%	0.1423%
48		18.02.t(1)	1706.35	-0.02%	-0.298%	0.702%	0.0000%
49		19.02.t(1)	1707.59	-0.07%	-0.296%	0.704%	-0.2845%
50		20.02.t(1)	1721.79	-0.83%	-0.297%	0.703%	0.1421%
51		21.02.t(1)	1717.67	0.24%	-0.297%	0.703%	0.0000%
52		24.02.t(1)	1706.52	0.65%	-0.298%	0.702%	0.1423%
53		25.02.t(1)	1703.43	0.18%	-0.299%	0.701%	0.1426%

I	J	K
Confidence Level	95.00%	95.00%
Alpha (α)	5%	5%
(1-α)-Quantile of Standard Normal Distribution	1.645	1.645
VAR = VaR Continuous Return, Portfolio	1.33%	1.33%
Portfolio Value in USD	10,000	10,000
VaR in USD, Portfolio	133	133
Time Period (N)	1	30
Time Factor	1.00	5.48
VaR Time Factor, continuous Return, Portfolio	1.33%	6.60%
VaR Time Factor, in USD, Portfolio	133	660
Density Function at point of (1-α)-Quantile	0.10313564	0.10313564
Conditional Value at Risk Time Factor	1.68%	8.49%
Conditional Value at Risk Time Factor, in USD	168	849

Fig. 6.4 Calculation of the Value at Risk and Conditional Value at Risk using the variance–covariance method

- Determine the VaR of the portfolio based on the previous assignment.

```
Alpha = 0.05;
pd = fitdist(Portfolio_Return,'Normal');
Norm_Distribution = norminv((1-Alpha));
Portfolio_volume = 10000;
T = 1;
Time_factor = sqrt(T);
VaR = Mean_Return+Portfolio_Std*Norm_Distribution % Alternative: icdf(pd,(1-Alpha));
VaR_USD = VaR*Portfolio_volume;
VaR_Time = Time_factor*Portfolio_Std*Norm_Distribution +T*Mean_Return;
VaR_Time_USD = VaR_Time*Portfolio_volume;
```

- Determine the CVaR of the portfolio.

```
Density = normpdf(Norm_Distribution);
CVaR_Time = Time_factor*Portfolio_Std*(Density/Alpha) +T*Mean_Return;
CVaR_Time_USD = CVaR_Time*Portfolio_volume;
```

Matlab Results (Figs. 6.5 and 6.6)

Literature and Software References

	VaR	VaR_USD	VaR_Time	VaR_Time_USD
1	0.0133	133.0780	0.0133	133.0780

Fig. 6.5 The calculated VaR in Matlab

Fig. 6.6 The calculated CVaR in Matlab

	CVaR_Time	CVaR_Time_USD
1	0.0168	167.5962

> Hull, J. (2018). Risk Management and Financial Institutions. 5th ed. Wiley. p. 317–324.
>
> See Excel file: `Case Study Risk Management`, Excel worksheet: `Variance-Covariance-Method`.
> See Matlab script: `A35_36_Var_Covariance`.

Assignment 37: Generation of Copulas

Task

For the aggregation of different loss distributions, we plan to use a Copula. First, generate the following Copulas to help you decide on a suitable Copula:

1. Gaussian Copula with $\rho = 0$
2. Gaussian Copula with $\rho = 0.9$
3. Gumbel Copula with $\lambda = 1$
4. Gumbel Copula with $\lambda = 4$
5. t-Copula with 3 degrees of freedom
6. t-Copula with 100 degrees of freedom

Can you think of situations in which situations you would use these Copulas?

Content

> Copulas are an increasingly popular method in finance. *Copulas* allow marginal distributions to be modelled and estimated separately from their

(continued)

dependence structure. Thus, they offer the possibility to use improved estimation methods for marginal distributions, since any assumptions about dependencies between assets are already covered by the Copula.

In mathematical terms, a *Copula* is a function that can specify a relationship between the *marginal distribution functions* of different random variables and their joint probability distribution. Copula functions can be used to combine arbitrarily distributed random variables with arbitrary dependence structures to form a new joint distribution function. There are different Copula functions. The most common Copulas are the Gaussian Copula (normal Copula), the t-Student Copula the Clayton Copula or the Gumbel Copula.

As an initial situation, we imagine d risks, each modelled as a random variable or represented by historical data. By calibration, we can determine for each risk the univariate distribution function. However, this knowledge alone is not sufficient to determine the joint distribution function of the risks, since we still lack all information about the dependence between the individual variables. An important structure for the determination of the dependence is provided by Sklar's theorem. This theorem states that every multivariate distribution function F can be split into

- a Copula C and
- the marginal distributions.

The marginal distributions are plugged used as arguments in the Copula to obtain the joint distribution.

Formally, the method is as follows:

- In a first step, the marginal distributions of the d risks are mapped to uniformly distributed $[0, 1]$-distributions. Typically, this is done using the corresponding quantiles of the marginal distributions (quantile mapping).
- In the second step, the dependency structure is chosen.

The main feature of a copula is that the multivariate cumulative distribution function described as the marginal distributions of the variables and their dependence structure can be determined separately from each other. This is very convenient in various situations, since it separates the often very complicated investigation of the joint distribution function into an investigation of the marginal distributions, on the one hand, and the Copula, on the other, in which all information about the dependence is encoded in the latter. Conversely, a (mathematically correct) multivariate distribution function can always be constructed from any marginal distributions and any Copula.

Despite the numerous possibilities of risk aggregation using Copula functions, we would like to point out the following critical issue. If the Copula function is chosen inappropriately for the corresponding problem (and the typically pre-specified marginal distributions), a mathematically correct joint

(continued)

> distribution can be generated, but this does not necessarily have to be economically meaningful or problem-adequate. Often, the most common Copulas (e.g., the Gaussian copula) are simply used without scrutinising their properties and implications for the current problem.

Important Formulas

Due to the complexity of Copula functions, we cannot cover this exciting topic comprehensively in this book. However, we want to convey the basic idea of Copulas and their importance for risk management.

The main theorem of the Copula is Sklar's theorem. Suppose we have distribution functions $F_1, F_2, \ldots, F_n$. Then the joint multivariate distribution function $F(x_1, x_2, \ldots, x_n)$ can be described as a Copula of these distribution functions:

$$F(x_1, x_2, \ldots, x_n) = C(F_1(x_1), F_2(x_2), \ldots, F_n(x_n)) \tag{6.6}$$

$F_1, F_2, \ldots, F_n$ = Distribution functions of the marginal distributions
F = Multivariate/joint distribution function
C = Copula function

Thus, Sklar's theorem states that a Copula allows the decomposition of a multivariate distribution function into a dependence structure and the marginal distributions.

For the marginal distributions holds:

$$u_1 = F_1(x_1), u_2 = F_2(x_2), \ldots, u_n = F_n(x_n) \tag{6.7}$$

$F_1, F_2, \ldots, F_n$ = Distribution functions of the marginal distributions

If the marginal distributions are continuous, the inverses $F_1^{-1}(u_1) = x_1, \ldots, F_n^{-1}(u_n) = x_n$ can be calculated. In this case the known representation for the Copula results as:

$$C(u_1, u_2, \ldots, u_n) = F(F_1^{-1}(u_1), \ldots, F_n^{-1}(u_n)) \tag{6.8}$$

$F_1^{-1}(\cdot), \ldots, F_n^{-1}(\cdot)$ = Inverse of the marginal distributions

In the following, we consider two-dimensional Copulas. For stochastic independent risks we have the following Copula:

$$C(u_1, u_2) = u_1 \cdot u_2 \tag{6.9}$$

u_1, u_2 = Marginal distributions

Assignment 37: Generation of Copulas

The normal or also Gaussian Copula is defined using the normal distribution function $N(\cdot)$:

$$C_\rho(u_1, u_2) = N_{2,\rho}\left(N^{-1}(u_1), N^{-1}(u_2), \rho\right) \qquad (6.10)$$

$N_2(\cdot, \cdot, \rho) =$ Bivariate distribution function of two standard normally distributed random variables
$\rho =$ Correlation coefficients

$N_2(\cdot, \cdot, \rho)$ is the bivariate distribution function of two standard normally distributed random variables with the correlation coefficient ρ.

Generating variables which are distributed according to the normal Copula with parameter $\rho = 0,5$ there is already a slight concentration of points along the bisector. For $\rho = 0$ the points are linearly independent.

The Gaussian Copula and the variance–covariance approach are based on the normal distribution. The normal distribution uses a mean value and a dispersion around the mean value (covariance matrix). With a normal distribution, values diverging far from the expected value occur only with a very small probability. That is, extreme events in which multiple losses occur simultaneously cannot be modelled with the variance–covariance approach and the Gaussian Copula. There, the probability of occurrence is underestimated. The assumptions of the normal distribution and linear dependencies are still satisfied to a certain extent when modelling market prices. This is not sufficient for other risk types such as credit risk or operational risk. For these risk types, extreme events should also be modelled. In Copula models, Sklar's theorem allows any univariate distribution with a dependence structure to be aggregated into a multivariate aggregate distribution. A well-known Copula for modelling extreme events is the Gumbel Copula.

The Gumbel Copula allows dependencies in the upper tails of the distributions. The figure displays a cluster of points in the upper right (upper tail dependence). The Gumbel Copula can be described as follows:

$$C_\lambda(u_1, u_2) = \exp\left(-\left((-\ln u_1)^\lambda + (-\ln u_2)^\lambda\right)^{\frac{1}{\lambda}}\right), \lambda \geq 1 \qquad (6.11)$$

$\lambda =$ Parameter of the Gumbel Copula

If you choose the parameter $\lambda = 1$ you have independence. The larger λ, the greater the dependence is modelled.

With the help of the Gumbel Copula, the dependency in the upper tail can be modelled. The t-Student Copula, on the other hand, allows dependencies to be modelled in the centre of the distribution as well as in the upper and lower tail.

$$C_v(u_1,u_2) = t_{\rho,v}\left(t_v^{-1}(u_1), t_v^{-1}(u_2), \rho\right) \qquad (6.12)$$

$t_{\rho,\,v}$ = t-distribution with correlation matrix ρ and degrees of freedom v
t_v^{-1} = The inverse of the univariate t-distribution with v degrees of freedom
ρ = Correlation matrix

Copulas cannot be modelled easily with Excel. Therefore, this application must be done, for example, in Matlab.

Execution in Matlab
- Define the assumptions given in the task.

```
n = 500; % Number of random variables
rho_low = 0; % Correlation factor
rho_high = 0.9;
lambda_low = 1; % Dependency parameter
lambda_high = 4;
nu_low = 3; % Degrees of freedom
nu_high = 100;
```

- Model the Gaussian, Gumbel and t-Student Copulas for high and low dependencies.

```
U_Gaus_h = copularnd('Gaussian',rho_high,n);
U_Gaus_n = copularnd('Gaussian',rho_low,n);
U_Gum_h = copularnd('Gumbel',lambda_high,n);
U_Gum_n = copularnd('Gumbel',lambda_low,n);
U_t_h = copularnd('t',rho_high,nu_high,n);
U_t_n = copularnd('t',rho_low,nu_low,n);
```

- Graphically represent the comparison of the Copulas.

```
t = tiledlayout(3,2,"TileSpacing","compact");
% Gaussian high
nexttile
plot(U_Gaus_h(:,1),U_Gaus_h(:,2),'.')
title('Gaussian {\it\rho} = 0.9')
% Gaussian low
nexttile
```

(continued)

```
plot(U_Gaus_n(:,1),U_Gaus_n(:,2),'.')
title('Gaussian {\it\rho} = 0')
% Gumbel high
nexttile
plot(U_Gum_h(:,1),U_Gum_h(:,2),'.')
title('Gumbel {\it\lambda} = 4')
% Gumbel low
nexttile
plot(U_Gum_n(:,1),U_Gum_n(:,2),'.')
title('Gumbel {\it\lambda} = 1')
% t-Student high
nexttile
plot(U_t_h(:,1),U_t_h(:,2),'.')
title('t-Student {\it\nu} = 100')
% t-Student low
nexttile
plot(U_t_n(:,1),U_t_n(:,2),'.')
title('t-Student {\it\nu} = 3')

title(t,'Copula Comparison')
xlabel(t,'X')
ylabel(t,'Y')
```

Matlab Results (Fig. 6.7)

Interpretation for the use of the Copulas: the Gaussian Copula with parameter $\rho = 0$ and the Gumbel Copula with parameter $\lambda = 1$ can be used to model independent risks. The t-Copula with $\nu = 100$ degrees of freedom approximates the Gaussian Copula. The Gaussian Copula, the traditional method for modelling dependence, mainly captures dependence in the middle of the distribution and implies independence in the tails, i.e. both very large and very small values at the tail of the distribution. The Gumbel Copula mainly captures dependence in the right tail of the distribution. This Copula has a strong concentration of probabilities near (0,0), so it correlates small losses. The t-Student Copula can capture dependence in the middle as well as in the tails of the distribution.

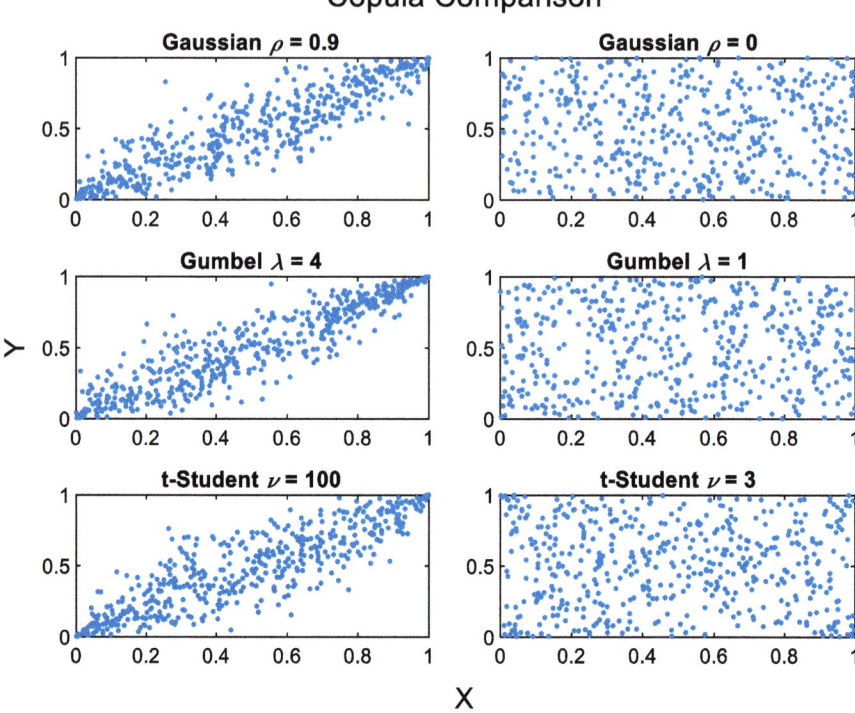

Fig. 6.7 The Gaussian, Gumbel and t-Student Copula for high and low dependencies

Literature and Software References

> Cont., R., Tankov, P. (2018). Financial modeling with jump processes. 2nd ed. London: Chapman & Hall/CRC (Chapman & Hall/CRC financial mathematics series), pp. 136–143.
> McNeil., A., Frey, R., Embrechts, P. (2015). Quantitative Risk Management. Concepts, Techniques and Tools, Princeton University Press, pp. 220–234.
> See Matlab script: `A37_Copulas`.

Assignment 38: Modelling the Aggregated Risk Using Copulas

Task
Consider the aggregated portfolio consisting of market and credit risks.

The assets of the market portfolio consist of 80% MSCI World Index equities and 20% bonds with coupon payments equal to 3-month EURIBOR +1%. The total sum invested is 1 million USD.

The exposure for the credit risk is also 1 million USD. Take the probability of default, the default correlation and the loss given default (LGD) from Assignment 3.6. The proportion between the exposure for market risk and credit risk is 50% to 50%.

Note: Proceed with the aggregation step by step:

1. First, aggregate the loss distributions of market risk using a Gaussian Copula.
2. Then aggregate the loss distribution of the market risk with that of the credit risk. Use a t-Copula for this purpose.

Calculate the Value at Risk and the Expected Shortfall. Also, determine the diversification effect by comparing this result to a situation where the risks are all independent of each other.

Content

The risks of banks and insurance companies consist of inter alia market, credit, insurance and operational risks. Changes in market parameters, such as interest rates or share prices, lead to a change in the value of assets. With respect to credit business, there is a risk that the debtor will not be able to repay the money borrowed. Other risks are of an operational nature and result, for example, from acts of fraud. In the previous assignments, we focused on modelling the individual risks. Ideally, the individual risks do not materialise simultaneously, but develop in opposite directions, so that the gain from one risk can pay for the loss from another risk. Often, therefore, the *total risk* is smaller than the sum of the individual risks. There are models that consider risk reduction through diversification when aggregating risks. For this purpose the variance–covariance method, as well as Copulas can be used.

It is worth mentioning that the Gumbel Copula holds a realistic property for aggregation due to the asymmetric marginal dependencies. However, the Gumbel copula is not suitable for aggregating more than two individual risks. There is only one parameter describing the dependence between the risk factors, therefore differences in pairwise dependencies cannot be considered. The t-Copula also has properties that allow for realistic modelling. With a t-Copula, more than two individual risks can be aggregated. However, it

(continued)

should be noted that the marginal dependencies can only be modelled symmetrically. In risk aggregation, therefore, not all individual risks are aggregated via a single copula, but the individual risks are aggregated into subgroups and combined with further risks at the next higher level in most cases.

Companies often use the aggregated risk to determine the *risk capital*. The risk capital describes the amount of capital that a company needs in order to be able to compensate (with a very high probability, often 99%) for losses realised within 1 year. This description can be expressed mathematically by the (mean) Value at Risk.

Execution in Matlab
- Import the required data and set the aforementioned assumptions.

```
Data = readmatrix('Matlab Data.xlsx');
MSCI = Data(:,3);
ThreeM_EURIBOR = Data(:,2);
Coupon = ThreeM_EURIBOR+1; % Coupon payment EURIBOR + 1%
Exposure_Market = 1000000; % 1 Mio USD
Exposure_Credit = 1000000; % 1 Mio USD
Weight = [0.8 0.2]; % Weight of stock and bonds in portfolio
PD = 0.01; % Annual Probability of Default
LGD = 1; % Loss Given Default
rho_Credit = 0.15; % Asset correlation
Risk_Division = [0.5 0.5]; % Division of exposure to market risks and credit risks
```

- To determine the market risk, calculate the negative continuous returns of the MSCI World and the bonds.

```
Return_Bond = -price2ret(Coupon); % Loss as positive value
Return_MSCI = -price2ret(MSCI); % Loss as positive value
```

- Visualise the frequencies of the returns as well as their correlation (Figs. 6.8 and 6.9).

```
t = tiledlayout(1,2,"TileSpacing","compact");
```

(continued)

Assignment 38: Modelling the Aggregated Risk Using Copulas

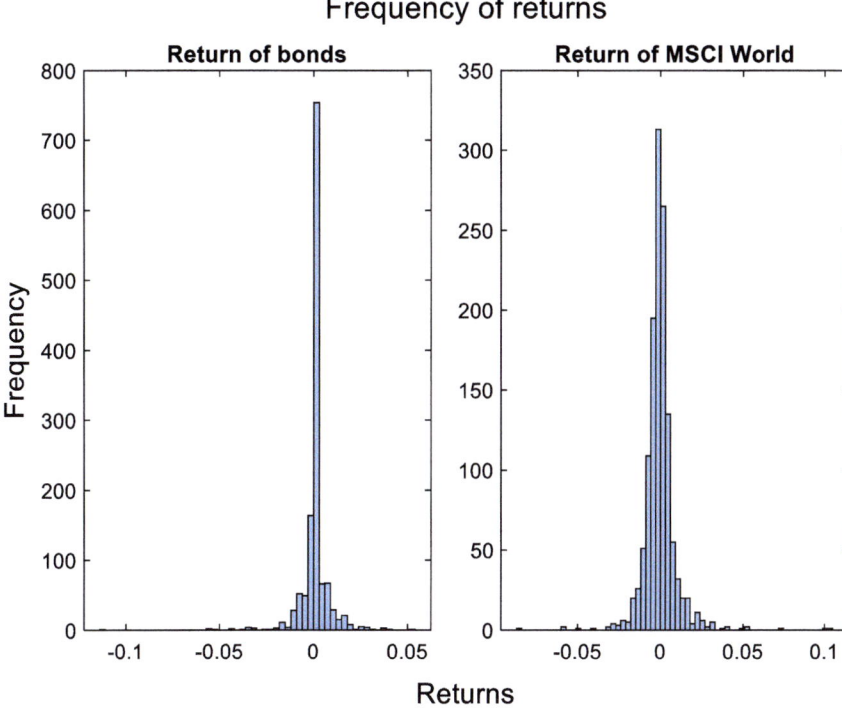

Fig. 6.8 Frequency of returns

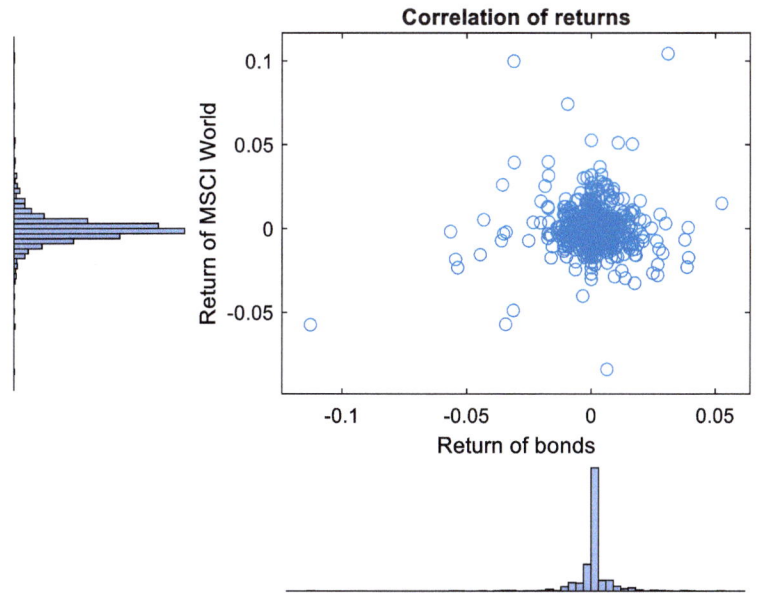

Fig. 6.9 Correlation of returns

```
nexttile
histogram(Return_Bond)
title('Return of bonds')

nexttile
histogram(Return_MSCI)
title('Return of MSCI World')

title(t,'Frequency of returns')
xlabel(t,'Returns')
ylabel(t,'Frequency')

figure
scatterhist(Return_Bond,Return_MSCI)
xlabel('Return of bonds')
ylabel('Return of MSCI World')
title('Correlation of returns')
```

- Estimate the probability densities based on the kernel function.

```
% Cumulated probability density(x,pts) of data x, assessed at points pts
k = ksdensity(Return_Bond,Return_Bond,'function','cdf');
m = ksdensity(Return_MSCI,Return_MSCI,'function','cdf');
M = [k m];
```

- Decisive for the determination of Copulas are their parameters. In Matlab these parameters can be either determined with the function copulafit() on the basis of real data or generated with the help of the function copulaparam().

```
rng default
Rho_hat = copulafit('Gaussian',M) % Fitting Gaussian Copula to returns
```

- Generate samples through the Gaussian Copula and plot the simulated returns (Fig. 6.10).

Assignment 38: Modelling the Aggregated Risk Using Copulas

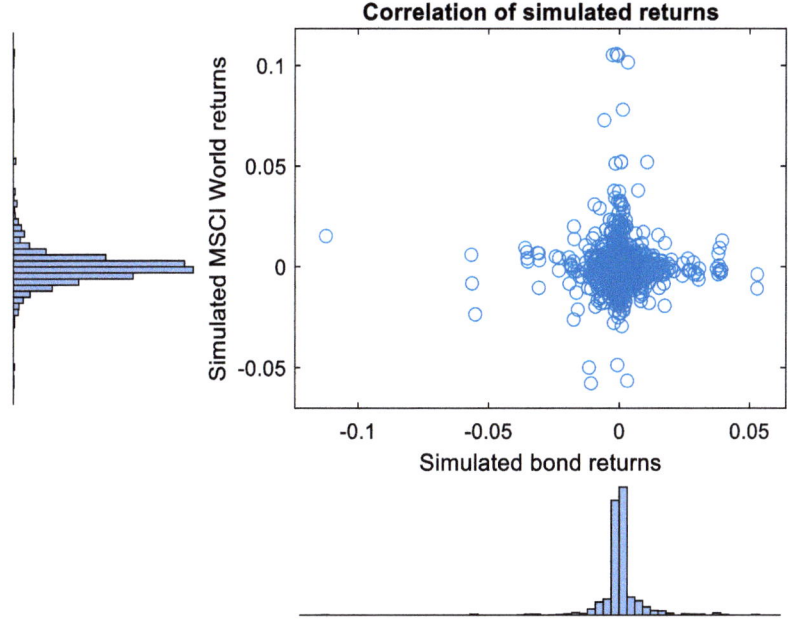

Fig. 6.10 Correlation of simulated returns

```
% Through Gaussian Copula simulated densities
G = copularnd('Gaussian',Rho_hat,length(Return_MSCI));
k1 = ksdensity(Return_Bond,G(:,1),'function','icdf'); % Simulated bond
returns in USD
m1 = ksdensity(Return_MSCI,G(:,2),'function','icdf'); % Simulated MSCI
World returns in USD
M1 = [k1 m1]; % Bond and MSCI World returns

figure
scatterhist(k1,m1)
xlabel('Simulated bond returns')
ylabel('Simulated MSCI World returns')
title('Correlation of simulated returns')
```

- Calculate the simulated returns of the portfolio.

```
% Weighted exposure to market risks in USD
Market_Exposure      =      (M1(:,1)*Weight(2)+M1(:,2)*Weight(1))
*Exposure_Market;
```

- To determine the credit risks, calculate the WCDR for an alpha between 0 and 1.

  x = [0.0000000001: ((0.2-0.0000000001)/(length(Market_Exposure)-1)): 0.2];
  pd = makedist('Normal'); % Probability distribution
  %cdf_Default = cdf(pd,((sqrt(1-rho)*icdf(pd,x)-icdf(pd,PD))/sqrt(rho))).'; % Cdf default rate
  Alpha = [0:1/1302:1]; % Possible alpha values between 0 and 1
  % Worst-Case-Default-Rate
  WCDR = (cdf(pd,((icdf(pd,PD)+sqrt(rho_Credit)*icdf(pd,Alpha))/sqrt(1-rho_Credit))));
  Credit_Exposure = (Exposure_Credit*LGD*WCDR).'; % Exposure to credit risks in USD

- To aggregate the risks, first visualise the relationship between the market and credit risks. Then estimate the probability densities of the market and credit exposures using the kernel function and fit a t-Student Copula to the densities (Fig. 6.11).

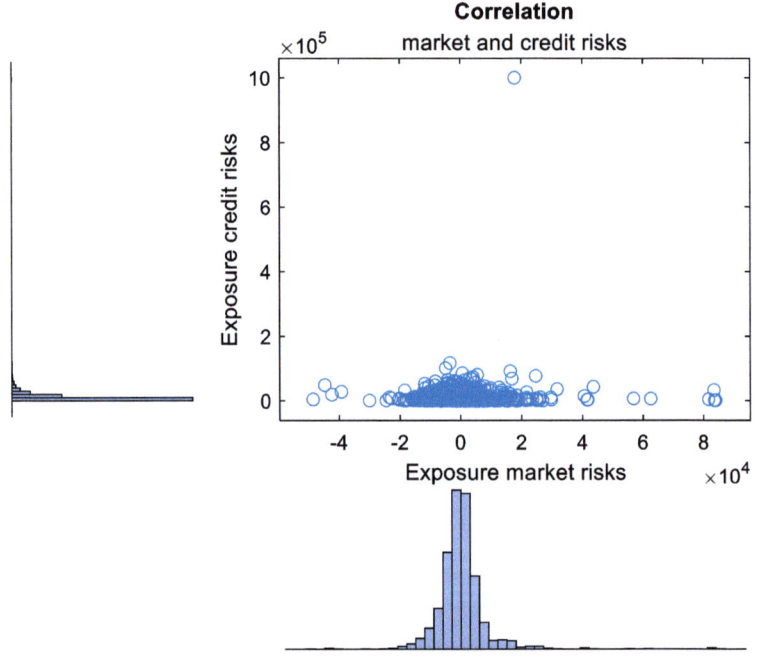

Fig. 6.11 Correlation of market and credit risks

```
figure
scatterhist(Market_Exposure,Credit_Exposure)
xlabel('Exposure market risks')
ylabel('Exposure credit risks')
title('Correlation', 'market and credit risks')
% Kernel-Density of the market risks
mr = ksdensity(Market_Exposure,Market_Exposure,'function','cdf');
% Kernel-Density of the credit risks
kr = ksdensity(Credit_Exposure,Credit_Exposure,'function','cdf');
R = [mr kr];
[Rho_hat2, nu_hat] = copulafit('t',R) % Fitting t-Student Copula to risks
```

- Generate samples through the t-Student Copula and plot them (Fig. 6.12).

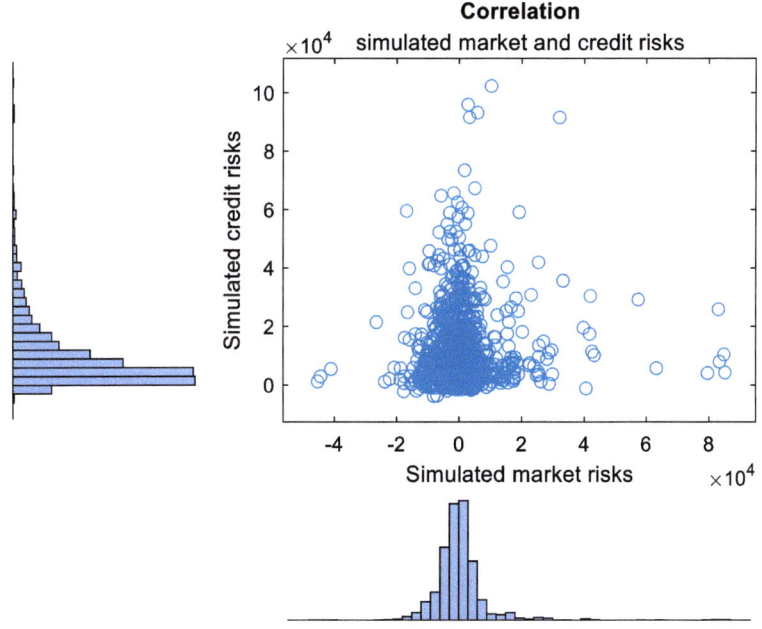

Fig. 6.12 Correlation of simulated market and credit risks

```
% Through t-Student Copula simulated densities
T = copularnd('t',Rho_hat2,nu_hat,length(Return_MSCI));
mr1 = ksdensity(Market_Exposure,T(:,1),'function','icdf'); % Portfolio market risks in USD
kr1 = ksdensity(Credit_Exposure,T(:,2),'function','icdf'); % Portfolio credit risks in USD

figure
scatterhist(mr1,kr1)
xlabel('Simulated market risks')
ylabel('Simulated credit risks')
title('Correlation', 'simulated market and credit risks')
```

- Calculate the simulated risk exposures.

```
Total_risk = mr1*Risk_Division(1)+kr1*Risk_Division(2); % 50:50 division of risk exposure in USD
```

- Calculate the Mean VaR and Expected Shortfall of the total loss distribution for an alpha of 1%.

```
VaR = quantile(Total_risk,0.99)-mean(Total_risk); % Mean VaR in USD for alpha = 0.01
ES = mean(Total_risk(Total_risk>VaR)); % Expected Shortfall in USD
```

- Calculate the diversification effect as the difference in risk exposure of diversified and independent risks.

```
% Undiversified risks
Total_risk_un = (Return_Bond*Weight(2)+Return_MSCI*Weight(1))*Exposure_Credit+WCDR.'*Exposure_Credit;
VaR_un = quantile(Total_risk_un,0.99)-mean(Total_risk_un); % Mean VaR
ES_un = mean(Total_risk_un(Total_risk_un>VaR_un)); % Expected Shortfall
Diversification_effect = mean(Total_risk_un)-mean(Total_risk); % + = Lower average losses
table([VaR; VaR_un],[ES;ES_un],'VariableNames',{'Mean Value at Risk (99%)', 'Expected Shortfall (99%)'},'RowNames',{'Aggregated risks', 'Sum of individual risks'})
table([Diversification_effect],'VariableNames',{'Diversification effect in USD per day'})
```

Matlab Results (Figs. 6.13 and 6.14)

		Mean Value at Risk (99%)	Expected Shortfall (99%)
1	Aggregated risks	31584.11	43468.60
2	Sum of individual risks	57912.60	116967.07

Fig. 6.13 Overview of risk measures for the aggregated risks and the sum of individual risks

	Diversification effect in USD per day
1	5147.82

Fig. 6.14 The diversification effect of aggregated risks

Literature and Software References

- Hull, J. Cont (2018): Risk Management and Financial institutions. 5th ed. Wiley finance series, pp. 319–320.
  McNeil., A., Frey, R., Embrechts, P. (2015). Quantitative Risk Management. Concepts, Techniques and Tools, Princeton University Press, pp. 220–234.
- See Matlab script: A38_39_Copulas.

Assignment 39: Risk Capital

Task
Calculate the risk capital for the company with the risks described in Assignment 6.4.

Content

The risks of banks and insurance companies are divided into market, credit, insurance and operational risks. In the previous assignments, we focused on modelling market and credit risks. In addition, in the last assignment, we introduced a way to aggregate the individual risks to determine the aggregated risk of a bank or insurance portfolio. This is the key parameter for determining

(continued)

the risk capital of a bank as well as the RAROC factor (risk adjusted return on capital).

Risk capital is intended to cover all risks to which a bank or insurance company is exposed to, therefore own funds should be kept available for this purpose. The risk capital should be able to cover unexpected losses with a certain level of security. The European supervisory board requires insurance companies to hold capital in the amount of the Mean Value at Risk at a safe level (confidence level) of 99.5% over the period of one year. As such that the probability of survival of the financial institution for the upcoming year is 99.5%. The capital adequacy regulations in banking (Basel) specify a safety level of 99.9% as the minimum capital requirement for approaches based on internal risk measurement. This minimum level corresponds to a BBB rating when the Basel II regulations were introduced.

The bank can then adjust its risk level so that the probability of survival corresponds to its target rating (cf. rating agency). For example, a minimum level of 99.97% is currently required for an AA rating (see Hull (2018), p. 586). For a BBB rating, the level is lower and is currently 99.8% (see Hull (2018), p. 586).

Execution in Matlab
- Calculate the risk capital as the Mean VaR at the 99.5% confidence level for a period of one year. For this, refer to the aggregated risks determined in the previous assignment and assume that one year comprises 252 trading days.

Risk_capital = (quantile(Total_risk,0.995)-mean(Total_risk))*sqrt(252);
% Mean VaR in USD for alpha = 0.005

Matlab Results (Fig. 6.15)

Fig. 6.15 Risk capital in USD

	Risk capital in USD
1	669030.68

Literature and Software References

Hull, J.C. (2018): Risk Management and Financial institutions. 5th ed. Wiley finance series, pp. 585–596.
McNeil., A., Frey, R., Embrechts, P. (2015). Quantitative Risk Management. Concepts, Techniques and Tools, Princeton University Press, pp. 220–234.

See Matlab script: A38_39_Copulas.

References

Artzner, P., Delbaen, F., Eber, J.-M., & Heath, D. (1999). Coherent measures of risk. *Mathematical Finance, 9*(3), 203–228. https://doi.org/10.1111/1467-9965.00068

BCBS. (2004). *International convergence of capital measurement and capital standards.* ISBN: 92-9197-669-5.

Black, F., & Scholes, M. (1973). The pricing of options and corporate liabilities. *Journal of Political Economy, 81*(3), 637–654. http://www.jstor.org/stable/1831029

Bloss, M. (2020). Financial engineering. In M. Bloss, M. Kleinknecht, & D. Sörensen (Eds.), *Financial engineering. Strategies, valuations and risk management. 4th, revised and expanded edition.* De Gruyter Oldenbourg. https://doi.org/10.1515/9783110659931/html

Bollerslev, T. (1986). Generalized autoregressive conditional heteroskedasticity. *Journal of Econometrics, 31*(3), 307–327. https://doi.org/10.1016/0304-4076(86)90063-1

Cont, R., & Tankov, P. (2015). *Financial modeling with jump processes* (2nd ed.). Chapman & Hall/CRC (Chapman & Hall/CRC financial mathematics series).

da Fonseca, J. S. (2020). Portfolio selection in euro area with CAPM and lower partial moments models. *Portuguese Economic Journal, 19*(1), 49–66. https://doi.org/10.1007/s10258-019-00153-4

Daldrup, A., & Institute of Information Systems. (2005). In M. Schumann (Ed.), *Kreditrisikomaße im Vergleich (Professur für Anwendungssysteme und E-Business - Publikationen Arbeitsbericht, 13/2005)*. Georg-August-Universität Göttingen. http://www.econbiz.de/archiv1/2010/101408_kreditrisiko_vergleich.pdf

Embrechts, P., Klüppelberg, C., & Mikosch, T. (1997). *Modelling extremal events for insurance and finance* (pp. 352–370). Springer.

Embrechts, P., Resnick, S. I., & Samorodnitsky, G. (1999). Extreme value theory as a risk management tool. *North American Actuarial Journal, 3*(2), 30–41. https://doi.org/10.1080/10920277.1999.10595797

Engle, R. F. (1982). Autoregressive conditional heteroscedasticity with estimates of the variance of United Kingdom inflation. *Econometrica, 50*(4), 987. https://doi.org/10.2307/1912773

Ernst, D., & Häcker, J. (2021). *Risk management in the company step by step* (1st ed.). UKV Verlag-Munich (utb).

Ernst, D., Häcker, J., Bloss, M., Dirnberger, M., Kleinknecht, M., & Plötz, G. (2016). *Financial Modeling* (2nd ed.). Schäffer-Poeschel Verlag. http://gbv.eblib.com/patron/FullRecord.aspx?p=4623611

Hull, J. (2011). *Risk management. Banks, insurance companies, and other financial institutions.* (2nd ed.: Munich [et al:]). Pearson Studium (wi - wirtschaft).

Hull, J. (2014). *Risk management: Banks, insurance companies and other financial institutions* (3rd ed.). Pearson Studium (wi - wirtschaft).

Hull, J. (2018). *Options, futures, and other derivatives* (9th ed.). Global Edition.

Hull, J. (2018). *Risk management and financial institutions* (5th ed.). Wiley (Wiley finance series).

Matlab Documentation: SDE Models, link: https://de.mathworks.com/help/finance/sde-objects.html#brg0w0_, Accessed 21.07.2021

Matlab Documentation: Simulating Interest Rates, link: https://de.mathworks.com/help/finance/example-simulating-interest-rates.html?searchHighlight=vasicek&s_tid=srchtitle, Accessed 21.07.2021

Merton, R. C. (1974). On the pricing of corporate debt: The risk structure of interest rates. *The Journal of Finance, 29*(2), 449–470. https://doi.org/10.2307/2978814

Merton, R. C. (1976). Option pricing when underlying stock returns are discontinuous. *Journal of Financial Economics, 3*(1-2), 125–144. https://doi.org/10.1016/0304-405X(76)90022-2

Nielsen, L. T. (1992). Understanding N (d1) and N (d2): Risk Adjusted Probabilities in the Black-scholes Model 1: INSEAD. http://www.ltnielsen.com/wp-content/uploads/understanding.pdf

Price, K., Price, B., & Nantell, T. J. (1982). Variance and lower partial moment measures of systematic risk: Some analytical and empirical results. *The Journal of Finance, 37*(3), 843–855. https://doi.org/10.2307/2327712

Romeike, F., & Hager, P. (2013). *Success factors risk management 3.0* (3rd ed., pp. 471–475). Springer Gabler.

Rüschendorf, L. (1996). *Distributions with fixed marginals and related topics*. JSTOR; Inst. of Mathematical Statistics. Lecture notes, monograph series/Institute of Mathematical Statistics, 28. http://www.jstor.org/stable/i397502

Shefrin, H. (Ed.). (2008a). *A behavioral approach to asset pricing. ScienceDirect (online service)* (2nd ed.). Academic Press/Elsevier (Academic Press advanced finance series).

Shefrin, H. (2008b). Irrational exuberance and option smiles. In H. Shefrin (Ed.), *A behavioral approach to asset pricing* (2nd ed., pp. 337–357). Academic Press/Elsevier (Academic Press ad-vanced finance series).

SPGlobal. (2020). Default, Transition, and Recovery: 2020 Annual Global Corporate Default And Rating Transition Study; https://www.spglobal.com/ratings/en/research/articles/210407-default-transition-and-recovery-2020-annual-global-corporate-default-and-rating-transition-study-11900573, accessed 5.8.2022.

Steinhoff, C. (2008). *Quantifizierung operationeller Risiken in Kreditinstituten*. Cuvillier Verlag.

Wewel, M. C., & Blatter, A. (2019). *Statistics in undergraduate business and economics. Methods, application, interpretation; with removable collection of formulas. 4th, updated reprint*. Pearson Studium (wi -wirtschaft).

Index

A
Aggregation, 183–209
ARCH model, 38–46
Autoregressive, 39

B
Backtesting, 149–154
Bar chart, 11
Black-Scholes, 68–92
Black swans, 172
Bond, 162
    fixed rate, 162
Brownian motion, 54

C
Calibration, 120
Clayton Copula, 193
Coherent risk measure, 181
Conditional tail expectation, 135
Conditional value at risk, 135–138, 145–149
    for a continuous distribution function, 145–149
Copula, 192–207
    Clayton, 193
    Gaussian, 193
    normal, 193
    t-Student, 193

Credit risk, 1
Convexity, 163
Coupon, 162

D
Debt securities, 161
Default correlation, 112
Delta, 78, 79
Density, 18, 19
Density function, 16–24
Diffusions parameter, 61
Distribution
    leptokurtic, 46
Distribution function, 16–24
    univariate, 193
Drift parameter, 61

E
Elasticity, 163
EWMA model, 30–38, 46
Expected default, 156
Expected shortfall, 135–138, 145–149, 173
Expected shortfall magnitude, 156
Expert survey, 120
Exposure at default (EAD), 115
Extreme value, 172–180
Extreme value theory, 172–180

F
Fat tails, 46, 173
Frequency distribution, 11

G
Gamma, 79
GARCH model, 45–53
Gaussian copula, 104, 193
Geometric Brownian motion, 53–60
Geometry
  fractal, 172
Greeks, 78–83

H
Heteroscedasticity, 39, 46
Histogram, 11–16
Hurdle, 154

I
Implied volatility, 83–86
Instantaneous return, 61
Insurance risk, 2
Itô's lemma, 56

L
Lag factor, 31
Liquidity risk, 2
Loss frequency, 120
Loss given default (LGD), 116
Loss ratio, 116
Loss severity, 120
Lower partial moments, 154–156
  first order, 154
  second order, 154, 158
  zero order, 154

M
Macaulay duration, 161–172
Market risk, 1
Markov property, 54
Maximum likelihood estimator, 31
Maximum likelihood method, 31, 46, 112
Mean default risk, 156
Mean reversion, 53
Mean reversion level, 61
Mean reversion process, 61
Mean value at risk (MVaR), 128, 132–135
Merton's model, 98–103

Method
  analytic, 183
  parametric, 183
Minimal capital requirement, 104
Modified duration, 161–172
Monotony, 181

N
Nominal interest rate, 162
Nominal value, 162
Normal copula, 193
Normal distribution, 17, 139, 173, 184

O
Operational risk, 1
Option, 69
Option price model, 69
Ornstein-Uhlenbeck prozess, 61

P
Pareto-Verteilung, 173
Peaks-over-Threshold method (PoT), 173
Poisson distribution, 120
Portfolio default rate, 107–111
Portfolio loss, 115–118
Portfolio risk, 183–188
Positive homogeneity, 181
Probability of default, 93, 103, 112
Put-call parity, 74–78

Q
Quantile, 127
Quantile mapping, 193

R
Random process, 53
RAROC factor, 208
Rating, 94
Rating agencies, 94
Rating migration matrices, 93–97
Reduction factor, 31
Return, 5–30
  continuous, 6
  discrete, 6
Return calculation, 5–11
Rho, 78, 80
Risk, 5
Risk capital, 207–209

Index

Risk measure, 127–154, 180–181
Risk ratios, 68–92

S
Shortfall expectation value, 154
Shortfall probability, 154–156
Shortfall variance, 154, 158
Sklar theorem, 193
Standard deviation, 27–30
Stochastic processes, 53–68
Strike price, 69
Student t-distribution, 17
Student-t-copula, 193
Subadditivity, 181

T
Tail VaR, 135–138
Target Shortfall, 156
Theta, 80
Threshold value, 156
Total risk, 199–207
Transalation invariance, 181

V
Value at risk, 127–132, 149–154
    conditional, 145–149
    for a continuous distribution function, 138–145
    for a discrete probability distribution, 127–132, 135–138
Variance, 24–27
    conditional, 46
    long-term, 46
Variance–covariance matrix, 183–188
Variance–covariance method, 183–192
Vasicek model, 61, 103–118
Vasicek/Ornstein-Uhlenbeck process, 61–68
Vega, 79
Volatility, 5–53
Volatility cluster, 38, 46
Volatility clustering, 38
Volatility smile/-surface, 86–92

W
Wiener process, 54
Worst Case default rate, 103–107

MIX
Papier aus verantwortungsvollen Quellen
Paper from responsible sources
FSC® C105338

If you have any concerns about our products,
you can contact us on
ProductSafety@springernature.com

In case Publisher is established outside the EU,
the EU authorized representative is:
**Springer Nature Customer Service Center GmbH
Europaplatz 3, 69115 Heidelberg, Germany**

Printed by Libri Plureos GmbH
in Hamburg, Germany